Ques

CB C

101 Que

HOWARD W. SAMS & CO., INC.
THE BOBBS-MERRILL CO., INC.
INDIANAPOLIS · KANSAS CITY · NEW YORK

SECOND EDITION

THIRD PRINTING—1975

International Standard Book Number: 0-672-20893-8
Library of Congress Catalog Card Number: 74-182875

Preface

This book is a handy guide and reference source for prospective and existing users of Citizens-band radio. It covers CB operating procedures, licensing, rules, and equipment selection. It is written in nontechnical language so as to make its contents understandable to everyone.

There are more than 800,000 Citizens radio licensees who are authorized to operate in excess of 5,000,000 transmitters—more than are operated in all other radio services combined. New applications for CB radio licenses are filed with the FCC at the rate of about 20,000 per month.

The Automobile Manufacturers Association is urging all motorists to install CB radios in their cars to enable them to summon aid when required. Motorists alone could use more than 90,000,000 CB transceivers.

A recent survey indicates that more than 43 percent of CB licensees use CB radio for business communications and that 36 percent of CB licensees are interested in a career in electronics.

CB licensees are required to possess a copy of Part 95, FCC Rules and Regulations, which may be purchased from the U.S. Government Printing Office for $1.75.

Leo G. Sands

Contents

PART ONE—Classes and Uses of CB Radio

PART TWO—Licensing and Rules

PART THREE—CB Operating Procedures

PART FOUR—CB Equipment

PART FIVE—Mobile and Fixed Installations

1

Classes and Uses of CB Radio

What is meant by Citizens band?

The "so-called" Citizens band is that portion of the radio spectrum between 26.96 MHz and 27.26 MHz. This band is divided into 23 channels which can be used for voice communication plus six channels which can be used for remote radio control of objects such as model aircraft and boats. There is also another Citizens band within the 450-470—MHz portion of the radio spectrum. However, the most popular Citizens band is the one in the 27-MHz range and is known for short as CB.

General information pertaining to the various classes in the Citizens Radio Service is summarized in Table 1-1.

What is the difference between a CBer and a ham?

A CBer is licensed to operate in the Citizens Radio Service. A ham is licensed to operate in the Amateur Radio Service. CBers are required to operate on specific channels only, whereas

Table 1-1. The Citizens Radio Service—General Information

Class of Station	Frequency Band (MHz)	Interstation Channels	Intrastation Channels	Normal Range (Miles)	Legal Range Limit (miles)	Equipment Cost	Use
A	460-470	16	16	10-40	None	$500-$1000	Voice (business and personal)
C	26.96-27.26 72-76	None for voice	None for voice	1-10	None	None for voice	Remote control only
D	26.96-27.26	7	23 (Including the 7 interstation channels)	1-5 (Walkie-talkies) 5-25 (Mobile-base) 10-35 (Fixed)	150	$6-$200 (Walkie-talkies) $70-$400 (Base and mobile)	Voice (business and personal)

hams are permitted to operate on any frequency within the bands authorized. A CBer is not permitted to use two-way radio as a hobby. But a ham is allowed to use two-way radio for "chit-chat" and for discussing the technical aspects of radio. A ham, however, is not allowed to use his set for business purposes.

In order to operate a CB transmitter, is it necessary to pass an examination?

Not as of now. However, the FCC has been urged to examine all CB license applicants on the applicable rules.

Why doesn't the FCC legalize hobby-type operation on the Citizens band?

FCC officials say that international agreements prohibit them from doing so. However, such agreements could be amended if the United States government would initiate discussions. Also, the Class-D CB category was created to provide a low-cost short-distance radiocommunication facility for business and personal use. This is inconsistent with hobby-type operation; with 800,000 licensees and only 23 channels hobby operations would create such congestion, few people would find the service adequate.

Since many CBers would become hams if it were not required that they pass a code test, why doesn't the FCC amend its rules and grant ham licenses without applicants having to demonstrate their capability to transmit and receive messages in International Morse Code?

International agreements allow the FCC to issue ham licenses to authorize voice transmission only on frequencies above 144 MHz without requiring applicants to pass a code test.

What is a Class-A Citizens radio station?

A Class-A station is one which operates on a specifically assigned channel in the 460-470—MHz uhf band. There are 16

Table 1-2. Class-A Citizens Radio Frequencies (in MHz)

462.550	462.600	462.675	467.550	467.625	467.700
462.575	462.625	462.700	467.555	467.650	467.725
	462.650	462.725	467.600	467.675	

channels available to Class-A stations (see Table 1-2). Transmitter power is limited to 60 watts input or 44 watts mean output power. Either a-m or fm may be used. The equipment for Class-A stations is considerably more expensive than for Class-D stations. Fig. 1-1 is a typical transceiver of the type which can be licensed as a Class-A Citizens radio station.

Courtesy Kaar Electronics Corp.

Fig. 1-1. Typical Class-A transceiver.

What is meant by a paired Class-A channel?

Class-A channels 462.550 MHz through 462.725 MHz may be used by mobile units only. When the base station transmits on 462.550 MHz, for example, and mobile units transmit on 467.550 MHz, the system operates in the two-frequency simplex mode. The base station receives from mobile units on 467.550 MHz and the mobile units receive from the base station on 462.550 MHz, as shown in Fig. 1-2. These two frequencies constitute a paired channel.

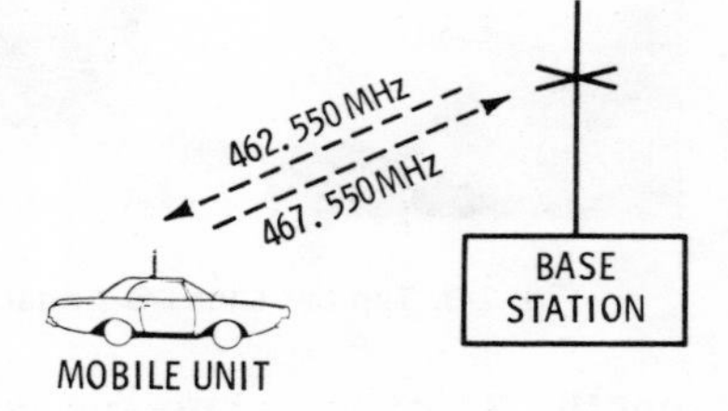

Fig. 1-2. A paired Class-A channel. Mobile units can communicate only with the base station, not mobile-to-mobile.

What is a Class-B Citizens radio station?

Class-B stations must go off the air when their licenses expire. Until the FCC stopped issuing new Class-B station licenses, a Class-B station could operate on 465 MHz using FCC type approved transceivers. The one shown in Fig. 1-3 made by Vocaline Corporation of America, was the first commercially successful CB transceiver. Also, under a Class-B license, operation was permitted on 48 other channels in the 460-470 MHz band when using equipment FCC type accepted for use by Class-A stations. At one time, the entire 460-470 MHz band was allocated to the Citizens Radio Service. There was room in that band for 400 channels. Later, the FCC authorized Citizens radio stations to operate on 49 channels within that band. More re-

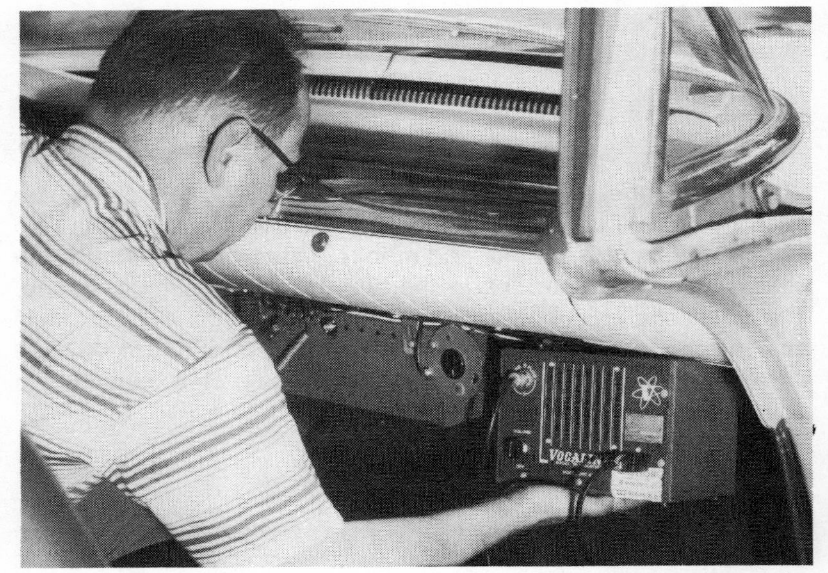

Photo by Jacques Saphier

Fig. 1-3. Typical Class-B transceiver formerly used on 465MHz only.

cently, 33 of the channels were reallocated to other services, leaving 16 channels for use by Class-A stations only and none for Class-B stations.

What is a Class-C Citizens radio station?

A Class-C Citizens radio station is one which is used for the remote control of objects such as model aircraft and ships. Such stations may also be used for signaling, but not for voice communications. There are six channels available in the 27-MHz Citizens band for Class-C stations as well as additional channels in the 72-76—MHz band.

What is a Class-D Citizens radio station?

Most CB radio stations are covered by a Class-D license. Class-D stations may operate on any of the 23 channels listed in Table 1-3 using a-m or ssb radiotelephony.

Table 1-3. Class-D CB Radio Channels

Channel Number	Frequency (MHz)	Channel Number	Frequency (MHz)
1	26.965	13*	27.115
2	26.975	14*	27.125
3	26.985	15*	27.135
4	27.005	16	27.155
5	27.015	17	27.165
6	27.025	18	27.175
7	27.035	19	27.185
8	27.055	20	27.205
9**	27.065	21	27.215
10*	27.075	22	27.225
11*	27.085	23*	27.255
12*	27.105		

*May be used for interstation communications.
**Emergency only.

What is Channel 22A?

Some CB transceivers were placed on the market which could be operated on Channel 22A (27.235 MHz) in anticipation of an FCC rules change which did not materialize. That channel may not be used unless the transceiver has been accepted for, and is licensed in, one of the Industrial, Land Transportation, or Public Safety radio services.

What are the advantages and disadvantages of Class-A CB operation compared to Class-D?

On the Class-A band there is no skip interference and ignition noise is seldom a problem. However, equipment cost is much higher.

What is a Class-E Citizens radio station?

At the time this book was published, the Class-E Citizens band had not yet been established by the FCC. Early in 1971, the Electronics Industries Association petitioned the FCC to reallocate the 220-222 MHz portion of the 220-225 MHz amateur band to the Citizens Radio Service for use by Class-E stations which would be authorized to use fm only. This new Citizens band would be divided into 80 channels.

What is a Class-D transceiver?

It may be a walkie-talkie, mobile, or fixed transceiver designed to operate on the 27-MHz CB channels. A typical Class-D transceiver is shown in Fig. 1-4.

Courtesy Anixter Bros.

Fig. 1-4. Class-D CB tranceiver operable on any of the 23 available channels.

What kind of range can be expected when using Class-D CB radio?

The range can vary from less than a mile, car to car, to as much as 25 miles from a base station to a car. Range is limited greatly by noise. The noise is mainly from other vehicles in the vicinity. Ignition systems radiate harmful radio interference. Terrain also has a great effect on range, as do tall buildings. Within Manhattan, for example, car-to-car range is quite limited because of the presence of so many tall, steel-framed buildings. In general, the ability to communicate 5 to 7 miles can be considered as average. And, when skip-transmission conditions exist, it is possible to communicate over distances of more than 1000 miles, but it is unlawful to do so.

What is a base station?

A base station is a radio station at a fixed location used primarily for communicating with mobile units (see Fig. 1-5). However, Class-D base stations are licensed as mobile units even if located at a fixed point.

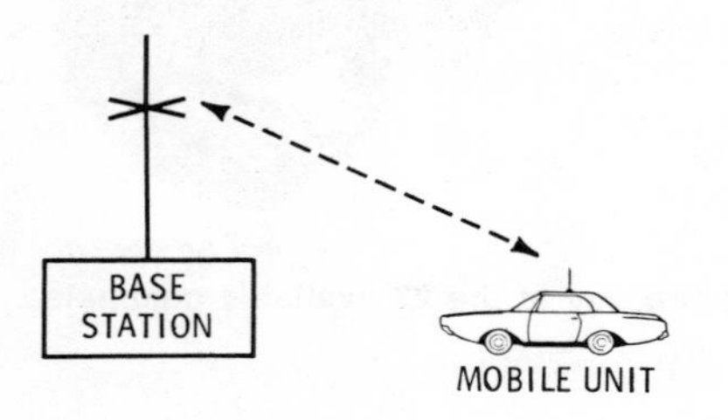

Fig. 1-5. Base station communicates primarily with mobile units.

What is a fixed station?

A fixed station is a radio station at a fixed location which is used primarily for communicating with other fixed stations.

Can CB radio be used for communicating between fixed locations?

Yes, CB radio transceivers can be used for point-to-point communication (see Fig. 1-6). It is permissible to communicate with stations of other licensees on any of the seven interstation channels (see Table 1-3). It is unlawful to do so on the 15 intrastation channels. The intrastation channels can be used for communicating between homes, offices, and other fixed loca-

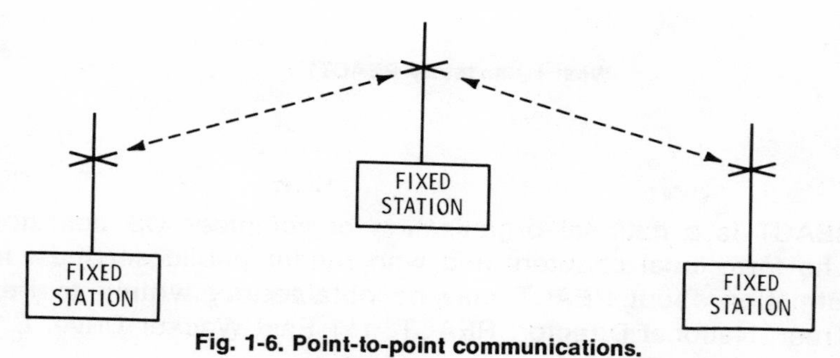

Fig. 1-6. Point-to-point communications.

tions when the transceivers at these locations are licensed to the same person and under his control.

Can CB operators participate in Civil Defense activities?

Yes. The way to go about it is to contact the local director of Civil Defense and volunteer your services. However, when using CB radio for official Civil Defense activities, communications must be conducted in accordance with FCC rules and regulations. A survey of Civil Defense organizations reveals that many of them use CB radio and encourage participation of individual CB operators.

What is HELP?

HELP means Highway Emergency Locating Plan. The intent of HELP is to make it possible for all motorists to use CB radio for obtaining aid in cases of emergencies on the road.

What is meant by REACT?

REACT is a national organization of volunteer CB operators who form local chapters and who render public services. Information about REACT may be obtained by writing to Pete Kreer, National Director, REACT, 111 East Wacker Drive, Chicago, Illinois 60601.

Is it permissible for an American CB station to communicate with a Canadian GRS station?

No, if the American CB station is within the United States. However, if the American CB station is within Canada and is covered by a Tourist Radio Permit, it is legal then to communicate with a Canadian station, licensed there in the General Radio Service.

What is the marine channel for CB communication?

No channel has yet been officially designated as a marine channel. However, Channel 13 is widely used as a ship-to-ship and ship-to-shore channel.

What is meant by REST?

REST stands for Radio Emergency Safety Teams, an organization of CBers which renders assistance to the pleasure boating public and which is dedicated to improving two-way radiocommunication between boats, yacht clubs, marinas, and shoreside CB mobile and base stations. REST teams operate both CB and vhf-fm base stations, so both CB and marine channels can be monitored. Information about REST membership can be obtained from Hugh F. Alexander, Executive Secretary, Radio Emergency Safety Teams, Inc., 300 Editors Building, 1729 H Street, NW, Washington, D.C., 20006.

Does the Coast Guard monitor CB channels?

No, not at the present time. Although CB Channel 13 has been designated as the unofficial marine watch channel, the Coast Guard does not monitor that channel. It is hoped that the government will allocate funds to the Coast Guard so that all Coast Guard stations can be equipped with CB transceivers.

Can CB radio be used on the high seas at a long distance from the American shore?

Yes, if there is anyone to communicate with. However, a CB station within the United States would not be permitted to communicate with a CB station on a boat which is more than 150 miles away. The British government has given permission to British ships to use CB radio when at sea and beyond interference range from the United Kingdom. The British government also allows British ships to use CB radio in foreign ports where the government of the foreign country has no objection.

Can CB radio be used on a personal airplane?

Yes. It is permissible to use CB radio equipment on any kind of aircraft (see Fig. 1-7). However, the use of walkie-talkies on

Courtesy Regency Electronics, Inc.

Fig. 1-7. A CB radio may be permanently installed on any aircraft.

commercial or other types of aircraft is prohibited when the aircraft is operated under instrument flying rules. No person, particularly a passenger, should attempt to operate a CB walkie-talkie on board a commercial airliner. The use of CB radio on private aircraft has many advantages. The owner may communicate over considerable distances with his home or office or with CB base stations at airports as well as CB mobile units.

2

Licensing and Rules

In order to use CB radio, is it necessary to have a license?

Yes, it is necessary to have a radio-station license issued by the Federal Communications Commission.

What does FCC stand for and where did the FCC get its right to regulate radiocommunications?

FCC stands for Federal Communications Commission. The Commission was created by an act of Congress which gives the Commission authority to regulate all radiocommunications within the United States. A copy of the Communications Act of 1934, as amended, is available from the Superintendent of Documents, U.S. Government Printing Office, Washington, D.C. 20402.

For how long is a CB radio-station license valid?

The license is usually valid for five years unless revoked earlier by the FCC for cause or change in regulations.

Do all countries require CB stations to be licensed?

No. Jamaica (West Indies), for example, permits operation of CB transceivers without a license.

In what countries is it illegal to use CB radio?

In all where the use of CB radio is not specifically authorized. In Italy, for example, the use of CB radio is unlawful, although it is said that there are thousands of Italian CBers, some of whom have been arrested.

Is it preferable to apply for a station license before buying equipment?

Yes. It is a good idea to apply for a license before buying equipment. When licenses are issued, the names and addresses of the licensees become public property, and various companies buy lists of names of CB licensees from organizations that gather this information. A new CB licensee usually receives literature and other information by mail describing equipment that is available to him. By the time he goes through this literature, he will have a better idea of what kind of equipment he wants to buy. Furthermore, it is usually very frustrating for a person to buy CB equipment and be unable to use it lawfully for the two to six weeks that usually lapse before his license is issued.

Who is eligible to apply for a Citizens radio-station license?

Any citizen of the United States over 18 years of age.

Is it necessary to get a special license to use CB radio for business communications?

No. A CB radio-station license authorizes the licensee to use two-way radio on CB channels for personal and business communication. However, many businesses may find it preferable to get a license in the Business Radio Service or some other radio service in which the applicant is eligible. In radio services other than Citizens and Amateur, personal communications are prohibited.

Is a special license required to use CB radio on a boat?

No. A CB radio licensee may install CB equipment at a fixed point or on any kind of a vehicle. However, on a boat operating on waters such as the St. Lawrence River, where the boat may cross the Canadian border, a CB licensee must obtain a Tourist Radio Permit from the Department of Transport of the Canadian government in order to permit him to use his CB equipment when on the Canadian side of the boundary.

Can a CB radio club get a radio-station license?

Yes, a club can obtain its own license, and its members can operate mobile units under that license, but only for the purpose of transacting club business. Members using CB radio for other than club business must have their own CB radio-station licenses.

Can an alien get a CB radio-station license?

No. Only a citizen of the United States is eligible to apply for a license in the Citizens Radio Service. However, an alien can use a transmitter which meets the requirements of Part 15, FCC Rules and Regulations, which cover the use of low-power unlicensed communications devices. However, an unlicensed station may not communicate with a licensed station.

How is a license obtained?

FCC Form 505 (Fig. 2-1) must be completed and signed and then mailed to the Federal Communications Commission, Gettysburg, Pennsylvania. It is necessary also to enclose a check or money order for $20.00 with the license application.

Instead of charging a $20.00 fee for a station license that can cover any number of CB transceivers, why doesn't the FCC charge a smaller fee for each transceiver and have the dealer collect the fee at the time of sale?

That has been suggested to the FCC, but the reply has been that administrative costs would be higher.

Where can license applications be obtained?

One is usually packed with every new CB transceiver. Copies of the form may also be obtained from any of the FCC field offices or by writing to the Federal Communications Commission, Washington, D.C. 20554.

On the license application form (505) there is no space for noting the number of stations to be used at fixed locations, only mobile units. Why is this?

All transmitters covered by a Citizens radio-station license (except Class A) are classed as mobile units, although they may be used at fixed locations as well as on vehicles. This means that the licensee may install and use CB transceivers at a number of unspecified, fixed locations.

FCC FORM 505
REVISED MAY 1963

UNITED STATES OF AMERICA
FEDERAL COMMUNICATIONS COMMISSION
WASHINGTON, D.C. 20554

FORM APPROVED
BUDGET BUREAU NO. 52–R123.10

APPLICATION FOR CLASS B, C, OR D STATION LICENSE IN THE

CITIZENS RADIO SERVICE

DO NOT WRITE IN THIS BLOCK

1. Application for class A station license must be filed on FCC FORM 400.
2. Complete on typewriter or print clearly.
3. Be sure application is signed and dated. *Mail* application to Federal Communications Commission, Gettysburg, Pa., 17325.
4. Enclose appropriate fee with application, if required. DO NOT SUBMIT CASH. Make check or money order payable to Federal Communications Commission. (See Part 19, volume VI of FCC rules to determine whether a fee is required with this application.)

1 NAME OF APPLICANT
BUSINESS NAME (OR LAST NAME, IF AN INDIVIDUAL)
FIRST NAME (IF AN INDIVIDUAL) | MIDDLE INITIAL

2 IF AN INDIVIDUAL OPERATING UNDER A TRADE NAME, GIVE INDIVIDUAL NAME; OR IF PARTNERSHIP, LIST NAMES OF PARTNERS (*Do not repeat any name used in item 1*)

LAST NAMES	FIRST NAMES	MIDDLE INITIAL

3 MAILING ADDRESS
NUMBER AND STREET
CITY | STATE
ZIP CODE | COUNTY OR EQUIVALENT SUBDIVISION

4 CLASSIFICATION OF APPLICANT (*See instructions*)
☐ INDIVIDUAL ☐ ASSOCIATION ☐ GOVERNMENTAL ENTITY
☐ PARTNERSHIP ☐ CORPORATION ☐ OTHER (*Specify*):

5 CLASS OF STATION (*Check only one*)
☐ CLASS B ☐ CLASS C ☐ CLASS D

6 IS THIS APPLICATION TO MODIFY OR RENEW AN EXISTING STATION LICENSE?
☐ YES (*Give call sign*): ☐ NO

7 DO YOU NOW HOLD ANY STATION LICENSE, OTHER THAN THAT COVERED BY ITEM 6, OF THE SAME CLASS AS THAT REQUESTED BY THIS APPLICATION?
☐ YES ☐ NO

8 TOTAL NUMBER OF TRANSMITTERS TO BE AUTHORIZED UNDER REQUESTED STATION LICENSE
(*Number*)

		YES	NO
9	DOES EACH TRANSMITTER TO BE OPERATED APPEAR ON THE COMMISSION'S "RADIO EQUIPMENT LIST, PART C," OR, IF FOR CLASS C OR CLASS D STATIONS, IS IT CRYSTAL-CONTROLLED? (*If no, attach detailed description: see subpart C of Part 19*)		
10	A. WILL APPLICANT OWN ALL THE RADIO EQUIPMENT? (*If no, answer B and C below*)		
	B. NAME OF OWNER		
	C. IS THE APPLICANT A PARTY TO A WRITTEN LEASE OR OTHER AGREEMENT UNDER WHICH THE OWNERSHIP OR CONTROL WILL BE EXERCISED IN THE SAME MANNER AS IF THE EQUIPMENT WERE OWNED BY THE APPLICANT?		
11	HAS APPLICANT READ AND UNDERSTOOD THE PROVISIONS OF PART 19, SUBPART D, DEALING WITH PERMISSIBLE COMMUNICATIONS FOR WHICH THIS CLASS OF STATION MAY BE USED?		
12	WILL THE USE OF THE STATION CONFORM IN ALL RESPECTS WITH THE PERMISSIBLE COMMUNICATIONS AS SET FORTH IN PART 19, SUBPART D?		
13	WILL THE STATION BE OPERATED BY ANY PERSON OTHER THAN THE APPLICANT, MEMBERS OF HIS IMMEDIATE FAMILY, OR HIS EMPLOYEES? (*If yes, attach a separate sheet listing the names and relationship of all such persons and give a detailed reason for their operation of your station*)		
14	IF APPLICANT IS AN INDIVIDUAL OR A PARTNERSHIP, ARE YOU OR ANY OF THE PARTNERS AN ALIEN?		
15	IS APPLICANT THE REPRESENTATIVE OF ANY ALIEN OR ANY FOREIGN GOVERNMENT? (*If yes, explain fully*)		
16	WITHIN 10 YEARS PREVIOUS TO THE DATE OF THIS APPLICATION, HAS THE APPLICANT OR ANY PARTY TO THIS APPLICATION BEEN CONVICTED IN A FEDERAL, STATE, OR LOCAL COURT OF ANY CRIME FOR WHICH THE PENALTY IMPOSED WAS A FINE OF $500 OR MORE, OR AN IMPRISONMENT OF 6 MONTHS OR MORE? (*See instructions. If yes, attach a separate sheet giving details of each such conviction*)		
17	IF APPLICANT IS AN INDIVIDUAL OR A PARTNERSHIP, ARE YOU OR ANY PARTNER LESS THAN 18 YEARS OF AGE (LESS THAN 12 YEARS OF AGE IF FOR CLASS C STATION LICENSE)?		

18 **IF THE PRINCIPAL LOCATION WHERE THE STATION WILL BE USED IS DIFFERENT FROM THE MAILING ADDRESS (ITEM 3), GIVE THAT LOCATION.** (*DO NOT GIVE POST OFFICE BOX OR RFD NO.*)
NUMBER AND STREET
CITY | STATE
IF LOCATION CANNOT BE SPECIFIED BY STREET, CITY, AND STATE, GIVE OTHER DESCRIPTION OF LOCATION

DO NOT WRITE IN THIS BOX
SCREENING ☐ Y ☐ N
SIGNATURE ☐ Y ☐ N

SIGN AND DATE THE APPLICATION ON REVERSE SIDE

Fig. 2-1. FCC Form 505—Application for Class-B, -C, or -D

19. IF APPLICANT IS A NONGOVERNMENTAL CORPORATION, ANSWER THE FOLLOWING ITEMS:		YES	NO
A	IS CORPORATION ORGANIZED UNDER LAWS OF ANY FOREIGN GOVERNMENT?		
B	IS ANY OFFICER OR DIRECTOR OF THE CORPORATION AN ALIEN?		
C	IS MORE THAN ONE-FIFTH OF THE CAPITAL STOCK EITHER OWNED OF RECORD OR MAY IT BE VOTED BY ALIENS OR THEIR REPRESENTATIVES, OR BY A FOREIGN GOVERNMENT OR REPRESENTATIVE THEREOF, OR BY ANY CORPORATION ORGANIZED UNDER THE LAWS OF A FOREIGN COUNTRY?		
D	IS APPLICANT DIRECTLY OR INDIRECTLY CONTROLLED BY ANY OTHER CORPORATION? *(If yes, answer items E through K below)*		
E	GIVE NAME AND ADDRESS OF CONTROLLING CORPORATION		
F	UNDER THE LAWS OF WHAT STATE OR COUNTRY IS THE CONTROLLING CORPORATION ORGANIZED?		
G	IS MORE THAN ONE-FOURTH OF THE CAPITAL STOCK OF CONTROLLING CORPORATION EITHER OWNED OF RECORD OR MAY IT BE VOTED BY ALIENS OR THEIR REPRESENTATIVES, OR BY A FOREIGN GOVERNMENT OR REPRESENTATIVE THEREOF, OR BY ANY CORPORATION ORGANIZED UNDER THE LAWS OF A FOREIGN COUNTRY? *(If yes, give details)*		
H	IS ANY OFFICER OR MORE THAN ONE-FOURTH OF THE DIRECTORS OF THE CONTROLLING CORPORATION AN ALIEN? *(If yes, answer items I and J below)*		
I	TOTAL NUMBER OF DIRECTORS IN CONTROLLING CORPORATION		
J	LIST ALL OFFICERS AND DIRECTORS WHO ARE ALIENS IN CONTROLLING CORPORATION AND GIVE BRIEF BIOGRAPHICAL STATEMENT FOR EACH ALIEN		

NAME	NATIONALITY	OFFICE HELD

K IS THE CONTROLLING CORPORATION IN TURN CONTROLLED BY OTHER COMPANIES? *(If yes, attach information for each of these controlling companies covering the information requested in items E through J, above)*

☐ YES

☐ NO

20. IF APPLICANT IS AN UNINCORPORATED ASSOCIATION, ANSWER THE FOLLOWING ITEMS:		YES	NO
A	IS ANY OFFICER OR DIRECTOR OF THE ASSOCIATION AN ALIEN?		
B	ARE MORE THAN ONE-FIFTH OF THE VOTING MEMBERS OF THE ASSOCIATION ALIENS OR REPRESENTATIVES OF ALIENS, FOREIGN GOVERNMENTS OR REPRESENTATIVES THEREOF, OR CORPORATIONS ORGANIZED UNDER THE LAWS OF A FOREIGN COUNTRY?		
C	IS THE ASSOCIATION DIRECTLY OR INDIRECTLY CONTROLLED BY ANY OTHER ORGANIZATION? *(If yes, give detailed explanation)*		

USE THIS SPACE FOR ANY ADDITIONAL INFORMATION OR REMARKS

WILLFUL FALSE STATEMENTS MADE ON THIS FORM ARE PUNISHABLE BY FINE AND IMPRISONMENT. U.S. CODE, TITLE 18, SECTION 1001.

ALL THE STATEMENTS MADE IN THE APPLICATION AND ATTACHED EXHIBITS ARE CONSIDERED MATERIAL REPRESENTATIONS, AND ALL THE EXHIBITS ARE A MATERIAL PART HEREOF AND ARE INCORPORATED HEREIN AS IF SET OUT IN FULL IN THE APPLICATION.

I CERTIFY THAT:

The applicant has (or has ordered from the Government Printing Office) a current copy of Part 19 of the Commission's rules governing the Citizens Radio Service;

The applicant waives any claim to the use of any particular frequency or of the ether as against the regulatory power of the United States because of the previous use of the same, whether by license or otherwise;

The applicant accepts full responsibility for the operation of, and will retain control of any Citizens Radio Station licensed to him pursuant to this application;

The station will be operated in full accordance with the applicable law and the current rules of the Federal Communications Commission

The said station will not be used for any purpose contrary to Federal, State or local law;

The applicant will have unlimited access to the radio equipment and effective measures will be taken to prevent its use by unauthorized persons; and

The statements in this application are true, complete, and correct to the best of my knowledge and belief and are made in good faith.

DO NOT OPERATE UNTIL YOU HAVE YOUR OWN LICENSE. USE OF ANY CALL SIGN NOT YOUR OWN IS PROHIBITED.

SIGNATURE: ______________________ DATE SIGNED: __________

(Check appropriate box below):

☐ INDIVIDUAL APPLICANT ☐ MEMBER OF APPLICANT PARTNERSHIP ☐ OFFICER OF APPLICANT CORPORATION OR ASSOCIATION ☐ OFFICIAL OF GOVERNMENTAL ENTITY

Citizens radio-station license.

Is it necessary to apply for a specific radio channel?

No. Under a Class-D Citizens radio-station license, the licensee is permitted to transmit on any of the 23 channels in the band. However, Class-A CB stations are authorized to operate only on one or more specifically assigned frequencies. The license applicant must specify the frequency on which he wishes to operate.

Is it necessary to obtain a license for every transceiver?

No. One license may cover any number of transceivers under the control of the license applicant.

How long does it take to get a CB radio-station license?

From the time the application is filed with the FCC in Gettysburg, Pennsylvania, it requires from two to six weeks before the license is issued. This waiting time may be reduced in the future since the FCC has a computer and the use of the computer may expedite the processing of applications. Since there are some 20,000 applications filed each month, it is reasonable to assume that there will be some time lag. The applications are read by human beings, not by machines, and some thought must be given to the acceptability of a license applicant.

Is it permissible to operate a CB transmitter pending issuance of a license?

No, it is not. Nor is it permissible to use someone else's license or the license of the dealer who sold the equipment.

Is it possible to get on the air on the Citizens band without having to wait for a license to be issued?

Yes. An application for a special temporary authority can be obtained by writing to the FCC giving the information listed as follows:

(1) Name, address, and citizenship status of applicant.
(2) Need for special action, including a description of any emergency or damage to equipment.
(3) Type of operation to be conducted.
(4) Purpose of operation.
(5) Time and date of operation desired.
(6) Class of station and nature of service.
(7) Location of station.
(8) Equipment to be used, specifying manufacturer, model number, and number of units.
(9) Frequency (or frequencies) desired.
(10) Plate power input to final radio-frequency stage.
(11) Type of emission.
(12) Description of antenna to be used, including height.

The FCC prefers to have at least 10 days' advance notice, but, under some circumstances, the FCC authorizes persons to operate radio transmitters by responding to a telegram requesting an STA (Special Temporary Authorization).

When the license arrives, what should be done with it?

The license should be posted at the main operating point of a CB radio system, which is usually the base station. When there is more than one base station, a photostatic copy of the license should be posted at all the stations. The original may be maintained in the licensee's files, if so desired.

What is meant by a *call sign?*

Every radio station is issued a *call sign.* CB stations are issued call signs which consist of three letters and four numerals—for example, KOD-1590.

What is a unit number?

It is the number designated by a CB radio licensee for one of his transceivers. A single CB radio-station license may cover any number of units, not exceeding the number designated on the station license. (Any applicant can apply for permission to operate any number of units.) The base station usually employs only its assigned call sign. All of the mobile units, which include walkie-talkies and transceivers installed in vehicles, have a unit number. For example, if the station license is KOD-1590, the mobile units must identify themselves as KOD-1590 Unit One, KOD-1590 Unit Two, etc.

What kind of identification should be provided at mobile units?

Each mobile transceiver should have affixed to it a copy of FCC Form 452-C (as revised). Fig. 2-2 illustrates this form. The form should be filled in to denote the call sign of the station license, the unit number, and the name and address of the licensee.

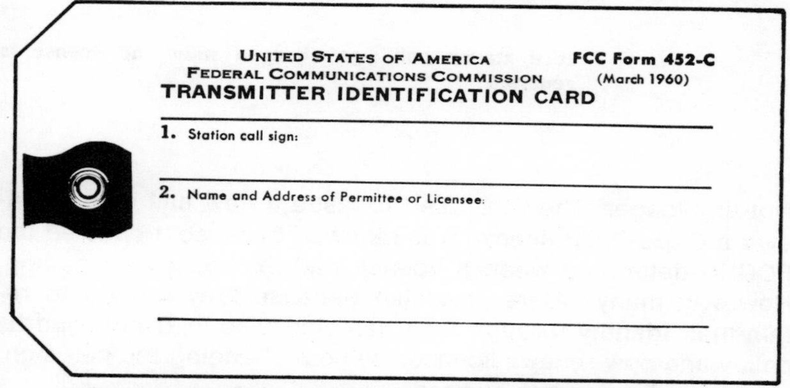
UNITED STATES OF AMERICA FCC Form 452-C
FEDERAL COMMUNICATIONS COMMISSION (March 1960)
TRANSMITTER IDENTIFICATION CARD

1. Station call sign:

2. Name and Address of Permittee or Licensee:

Fig. 2-2. FCC Form 452-C—Transmitter identification card.

If a CB licensee moves, is it necessary to apply for a new license?

No. It is necessary only to write a letter to the FCC in Gettysburg, Pennsylvania, stating that the licensee has moved, and his letter should contain the call sign, the old address, and the new address. This is not true of Class-A CB stations. When a Class-A CB station is moved, a modification of license application must be filed.

Can a CB radio-station license be renewed?

Yes, by filing FCC Form 505. However, a license-renewal application may be rejected by the FCC if the licensee has been cited for flagrant violation of the FCC rules.

Is a station call sign changed when the license is renewed?

Not any longer. The FCC used to issue a new call sign whenever a Class-D CB license was renewed because it enabled the FCC to determine when a license had expired by its coding. However, many CBers protested because they wanted to retain their identity through their call sign. The FCC changed its policy and now renews licenses without changing the call sign, unless the license has expired.

Can a CB radio-station license be transferred?

No. The license is issued to a specific person, corporation, or unincorporated association. It cannot be sold or transferred. However, if the ownership of a corporation is changed, it is necessary to file FCC Form 703, available from the FCC, Washington, D.C. 20554.

Is a license required for a walkie-talkie?

A Part-15 walkie-talkie, operated with power input under 100 milliwatts, may be used without a license for communication with other unlicensed stations. A walkie-talkie whose power input is greater than 100 milliwatts, such as the one shown in

Fig. 2-3. 3-channel walkie-talkie that may be used for both a-m and ssb transmission and reception, and which must be licensed.

Courtesy Midland International Corp.

Fig. 2-3, must be covered by a Class-D Citizens radio-station license. A Part-15 type walkie-talkie may be licensed, if it meets Class-D technical standards, to permit communicating with licensed stations.

What is a Part-15 walkie-talkie?

Under Part 15, FCC Rules and Regulations, it is permissible to operate a low-power communications device without a station license. Most of the low-cost walkie-talkies (see Fig. 2-4) which operate within the Citizens band are Part-15 types. Their power input is less than 100 milliwatts, and they may be operated on

Courtesy Allied Radio Shack

Fig. 2-4. Typical Part 15 Class-D walkie-talkie which does not require a station license.

any frequency within the 27-MHz Citizens band, not necessarily on a specific channel. When a Part-15 walkie-talkie is not covered by a CB radio-station license, it is unlawful to communicate with licensed CB stations.

What kind of license form is used for applying for a Class-A CB radio-station license?

FCC Form 400 is used for this purpose. The form is illustrated in Fig. 2-5. It is necessary to specify the location of the base station on this application form. The same application can cover any number of mobile units associated with the base station. However, a separate license application must be submitted for each station operated at a fixed location.

Where can I get a copy of the FCC Rules and Regulations covering the Citizens Radio Service?

All CB licensees are required to have a copy of the rules for reference, and these may be purchased from the U.S. Government Printing Office, Washington, D.C. 20402, for $2.00.

Does the FCC monitor CB radio transmissions?

Yes. The FCC has 18 monitoring stations as well as mobile monitoring stations that are dispatched to various areas. These monitoring stations are manned by personnel who measure frequencies of radio transmissions and monitor the operating habits of the operators. If violations are noticed, the FCC will send notices of violations. Failure to answer violation notices can result in cancellation of the license.

APPLICATION FOR RADIO STATION AUTHORIZATION IN THE
SAFETY AND SPECIAL RADIO SERVICES
Refer to "Instructions for Completion of FCC Form 400 (March, 1962 Revision)" and FCC Rules Pertaining to Particular Service
DETACH WORK SHEET, FILL OUT IN PENCIL, COMPLETE REMAINING SHEETS ON TYPEWRITER AND SUBMIT, WITH CARBONS ATTACHED, TO
FEDERAL COMMUNICATIONS COMMISSION, WASHINGTON 25, D. C.

FCC Form 400
March 1962

United States of America
FEDERAL COMMUNICATIONS COMMISSION

Form Approved
Budget Bureau No. 52-R132.7

This authorization permits the use of only such transmitters as are specified under "Special Conditions" and those appearing in the Commission's "Radio Equipment List, Part C" and designated for use in the particular radio service named in Item 4(a) of this application.

1(a). Frequency (Mc)	1(b). No. of transmitters			1(c). Emission	1(d). Maximum permissible power input (watts)
	Base or Land	Mobile	Other		

2(a). Name (see instructions)

(b). Mailing address (number, street, city, zone, state)

3. Location of transmitter(s) at a fixed location

Number and street (or other indication of location)

City	County	State

Latitude ° ' " N	Longitude ° ' " W

4(a). Name of radio Service

(b). Class of station:
Base ☐ Mobile ☐
Other (Specify):

COMMISSION FILE COPY
FOR COMMISSION USE ONLY

CALL SIGN

FILE NUMBER

5. If mobile units, or other class of station at temporary locations, are included in this authorization, show area of operation

6. Location of control point(s)

7. Overall height above ground of tip of antenna feet.

FOR COMMISSION USE ONLY

Antenna painting and lighting specifications:

Special Conditions:

This authorization effective __________

and will expire 3:00 AM EST. __________

and is subject to further conditions as set forth on reverse side. If the station authorized herein is not placed in operation within eight months this authorization becomes invalid and must be returned to the Commission for cancellation unless an extension of completion date has been authorized.

By direction of the FEDERAL COMMUNICATIONS COMMISSION

Secretary

8. Is applicant a representative of an alien or of a foreign government? If answer is "Yes", explain on the reverse of this page. Yes ☐ No ☐

9. State whether applicant is (Check one)
Individual ☐ Partnership ☐ Association ☐ Corporation ☐ Governmental Entity ☐
(If applicant is a non-governmental corporation fill out Item 19; if an unincorporated association fill out Item 20, on the reverse side of this page.)

10. If applicant is an individual, is applicant a citizen of the United States? Yes ☐ No ☐
If applicant is a partnership, are all partners citizens of the United States? Yes ☐ No ☐

11. Is communication service to be received from or rendered to another person (see instructions)? If "Yes", name of person is Yes ☐ No ☐

12.(a)(1) Will applicant own the radio equipment? Yes ☐ No ☐
If answer is "No", give name of owner

(a)(2) If not the owner of the radio equipment, is applicant a party to a lease or other agreement under which control will be exercised in the same manner as if the equipment were owned by the applicant? Yes ☐ No ☐

(b) Will applicant have unlimited access to the equipment and will effective measures be taken to prevent use of the radio equipment by unauthorized persons? Yes ☐ No ☐

13. Attach functional system diagram showing details of proposed radio system and include such other supplementary data as required by specific rules.

14. If it is proposed to use a transmitter which does not appear on the Commission's "Radio Equipment List, Part C" or if the transmitter is listed but not designated for use in the particular radio service named in Item 4(a) of this application, describe such transmitter in detail. (See Instructions)

15. Statement of eligibility

(Use space on the reverse of this page)

WORK SHEET

16.(a) Application for (Check one)
New station ☐ Assignment of authorization ☐
Modification ☐
Reinstatement of expired authorization ☐ License to cover CP ☐

(b) If for modification, state modification proposed

FOR COMMISSION USE ONLY

(c) If this application refers to a presently authorized station, give call sign

(d) Give points of communication (call signs)

(e) Are you presently authorized for any other stations in the service indicated in Item 4(a)? Yes ☐ No ☐

17. If antenna will be mounted on an existing antenna structure
(a) Give name of a licensee using this structure, his call sign and radio service and the current painting and lighting specifications required by the Commission for this antenna structure.

(b) If your proposed antenna will increase the height of the existing structure, give overall height above ground of the tip of the proposed antenna structure.

18.(a) Will the antenna extend more than 20 feet above the ground or natural formation, or more than 20 feet above an existing man-made structure (other than an antenna structure) on which it will be mounted? Yes ☐ No ☐

(b) Give height above ground for each component of the antenna structure (antenna, pole, tower, water tower, mast, building, chimney, etc., or combination of these) feet.

(c) Distance and direction to nearest aircraft landing area

(d) Elevation of ground above mean sea level at antenna site feet.

All the statements made in the application and attached exhibits (to , inclusive) are considered material representations, and all the exhibits are a material part hereof and are incorporated herein as if set out in full in the application.
The applicant certifies that he has a current copy of the Commission's Rules governing the radio service named in Item 4(a) above.
The applicant waives any claim to the use of any particular frequency or of the ether as against the regulatory power of the United States because of the previous use of the same, whether by license or otherwise, and requests an authorization in accordance with this application.
The statements in this application are true, complete, and correct to the best of my knowledge and belief and are made in good faith.

APPLICATION MUST BE SIGNED AND DATED

Applicant (Must agree with name as shown in Item 2(a)). (Date)

By __________

(Designate appropriate classification below)

☐ Individual Applicant
☐ Officer of Applicant Corporation or Officer and Member of Applicant Association
☐ Member of Applicant Partnership
☐ Official of Governmental Entity Competent under the Jurisdiction to Sign for the Applicant

WILLFUL FALSE STATEMENTS MADE ON THIS FORM ARE PUNISHABLE BY FINE OR IMPRISONMENT. U.S. CODE, TITLE 18, SECTION 1001.

(OVER)

Fig. 2-5. FCC Form 400—Application for radio station authorization in the

CONDITIONS OF GRANT

A. Subject to the provisions of the Communications Act of 1934, as amended, subsequent acts, treaties, and all regulations heretofore or hereafter made by this Commission, and further subject to the conditions and requirements set forth in this authorization the licensee or permittee hereof is authorized to use and operate the radio transmitting facilities herein described. This authorization shall not vest in the licensee or permittee any right to operate the station nor any right in the use of the frequencies designated in the authorization beyond the term hereof, nor in any other manner than authorized herein.

B. Neither this authorization nor the right granted herein shall be assigned or otherwise transferred to any person, firm, company, or corporation except by specific authorization of the Commission.

C. This authorization is issued on the licensee's representation that the statements contained in licensee's application are true and that the undertakings therein contained, so far as they are consistent herewith, will be carried out in good faith. The licensee shall, during the term of this license, render such service as will serve public interest, convenience, or necessity to the full extent of the privileges herein conferred.

D. This authorization is subject to the right of use or control by the Government of the United States conferred by Section 606 of the Communications Act of 1934, as amended.

FOR COMMISSION USE ONLY

19. If applicant is a non-governmental corporation

(a) Under the laws of what State is it organized?

Is more than one-fifth of the capital stock of the corporation either owned of record or may it be voted by aliens or their representatives or by a foreign government? Yes ☐ No ☐

Is any officer or director of such corporation an alien? If so, state name and position of each Yes ☐ No ☐

(b) Is applicant directly or indirectly controlled by any other corporation? If so, what is the name and address of the controlling corporation? Yes ☐ No ☐

Under the laws of what State is it organized?

Is more than one-fourth of the capital stock of the controlling corporation either owned of record, or may it be voted by aliens, their representatives, or by a foreign government or representative thereof, or by any corporation organized under the laws of a foreign country? Yes ☐ No ☐

Is any officer or more than one-fourth of the directors of such corporation an alien? If "Yes", state name and position of each, and state total number of directors Yes ☐ No ☐

Is the above described controlling corporation in turn a subsidiary? If so, attach additional sheets answering the items in this paragraph for each company to and including the organization having final control. Yes ☐ No ☐

20. If applicant is an unincorporated association

Number of members	Are any members aliens? Yes ☐ No ☐	Number of aliens, if any

Name and position of alien officers or directors, if any

21. Have you or any party to this application ever been convicted of any crime for which the penalty imposed was a fine of $500.00 or more, or an imprisonment of six months or more? Yes ☐ No ☐
If "Yes", you must attach a separate sheet with this form giving details of each such conviction.

REMARKS AND ADDITIONAL DATA

USE ADDITIONAL SHEETS IF NECESSARY

Safety and Special Radio Services including the Class-A Citizens band.

Does the public have anything to say about FCC rules and regulations?

Yes, indeed. Any citizen can file a petition to amend the rules or make new rules. The petition can be summarily dismissed by the FCC without action. When the FCC acts on a petition, a "notice of proposed rulemaking" is published in the Federal Register and citizens may then file their comments.

3

CB Operating Procedures

Who is permitted to operate a CB transceiver?

The licensee and members of his immediate family (see Fig. 3-1). Also, employees of the licensee may operate the licensee's CB radio equipment when performing their duties as employees. They may not use CB radio for any other purpose unless they have their own CB radio-station licenses.

How is the transmitter turned on and off?

Most CB transceivers are provided with a hand-held microphone such as the one shown in Fig. 3-2, which has a push-to-talk switch. To transmit, simply hold the microphone close to the mouth and press the switch with a finger, holding the switch in while talking. To receive, simply release the pressure on the switch. When the switch is pushed in, a relay in the transceiver disables the receiver and turns on the transmitter. When the

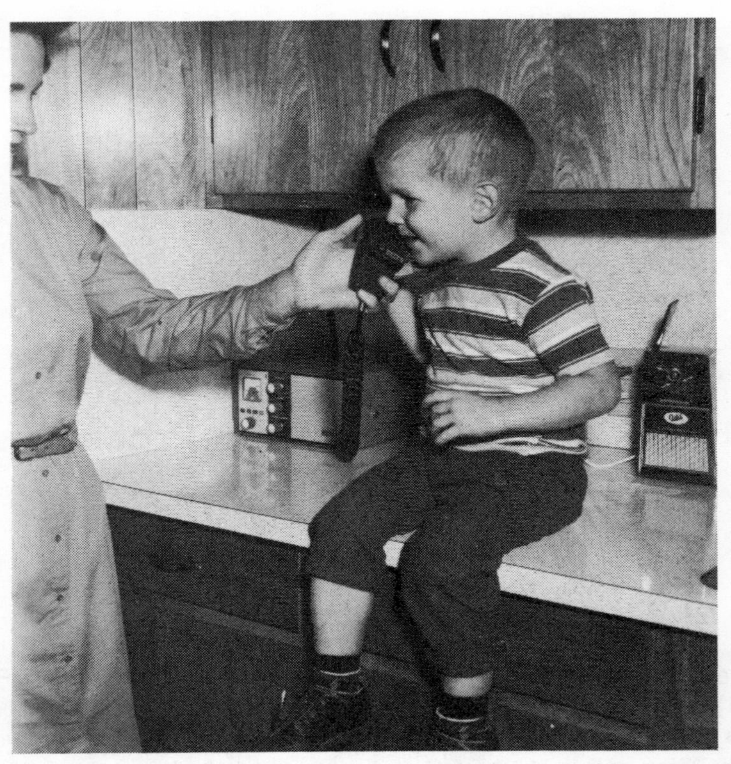

Fig. 3-1. Small children may use CB radio only under an adult's supervision.

switch is released, the transmitter is turned off and the receiver is reactivated.

What are the functions of the various controls of a CB transceiver?

Fig. 3-3 shows the front panel of a CB transceiver with several accessory features. The functions of the various controls are noted in the illustration. Not all transceivers have as many controls. Some have only volume, squelch, and channel-select controls.

Fig. 3-2. Hand-held microphone with push-to-talk switch at upper left.

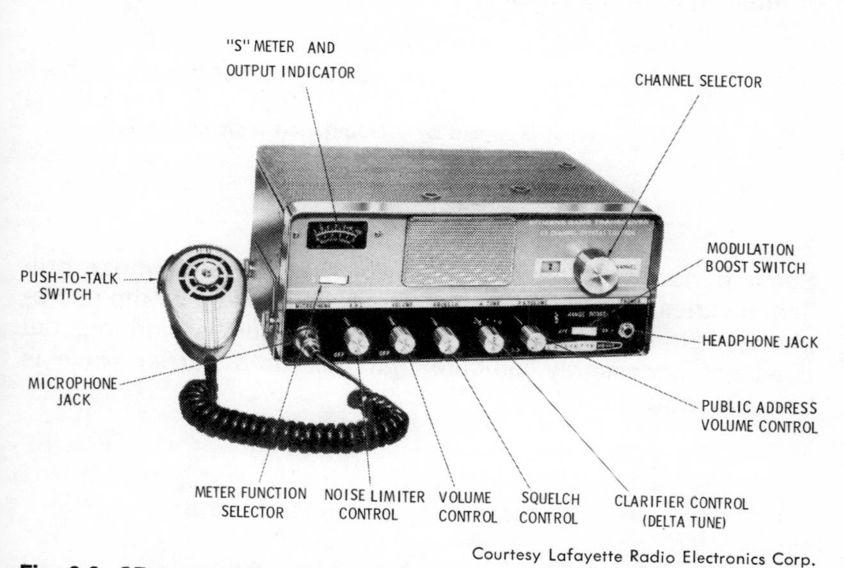

Courtesy Lafayette Radio Electronics Corp.

Fig. 3-3. CB transceiver showing functions of controls and other features.

What is a squelch control?

Without a squelch circuit, noise would be heard in the speaker when not receiving a radio signal. The squelch circuit silences the speaker when there is no radio signal. The squelch control is used for setting the squelch so that only signals stronger than a specific level will "awaken" the squelch and allow them to be heard.

What is a delta-tune control?

Some CB transceivers include a delta-tune control which enables the user to receive more clearly stations which are slightly off their authorized channel frequency.

What is meant by a modulation indicator lamp?

Some transceivers have a lamp which glows and whose brilliance varies with the level of the sound picked up by the microphone. It indicates that the modulator circuit is working, but does not necessarily indicate that the radio carrier wave is being modulated.

Should the engine of a vehicle be kept running when using a CB transceiver?

Not necessarily. However, with the engine running, the voltage across the vehicle battery is higher than when the engine is not running. With higher input voltage to the transceiver, transmitter output and receiver sensitivity usually increase.

Fig. 3-4. Right way to hold microphone.

Hold the microphone about 2 inches from the mouth and speak at a normal level. Holding the microphone farther away causes undermodulation and reduces the range. Right and wrong microphone techniques are illustrated in Figs. 3-4 and 3-5.

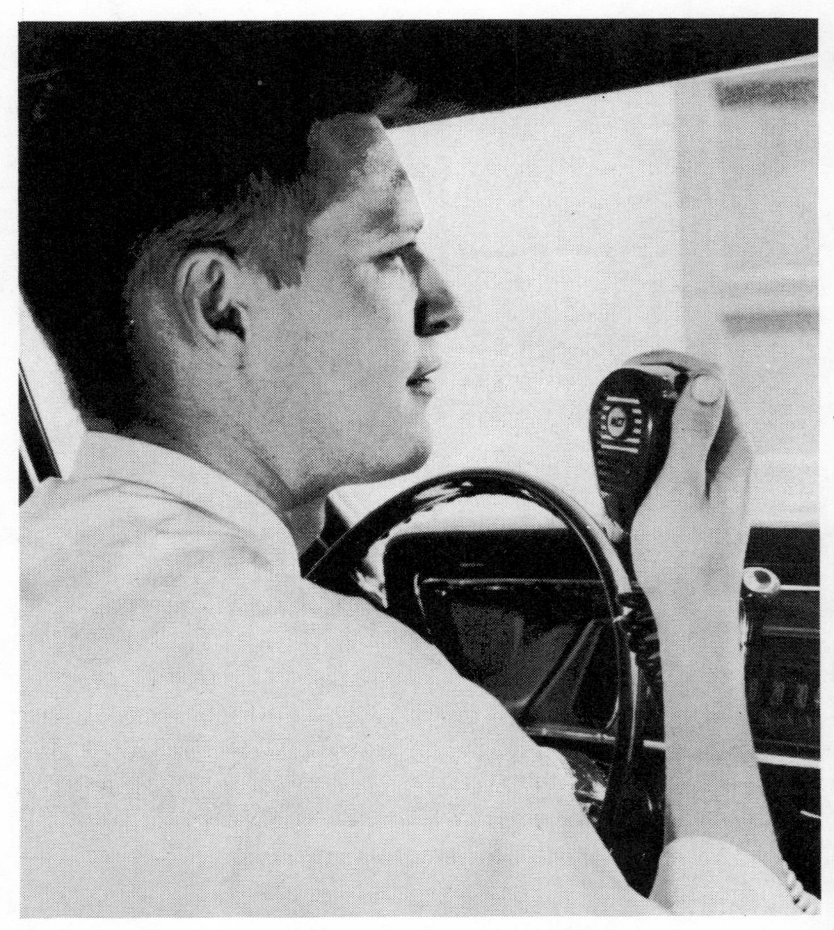

Fig. 3-5. Wrong way to hold microphone.

It is necessary to identify the station when transmitting?

It is necessary to identify the station at the beginning and end of every exchange of communication. It is also necessary to identify the station being communicated with. Identification means the call sign, and when applicable, the unit number in addition to the call sign.

Can a CB station communicate with any other CB station on any of the 23 Class-D channels?

No. Only eight of the channels are available for interstation communication. Of these, Channel 9 is restricted to emergency communications. The others are restricted to intrastation communication. Table 1-3 lists the Class-D channels and denotes which are available for interstation communications.

What is the procedure for initiating a call to another station?

Set the receiver to the desired channel and listen to make sure that the channel is clear. Then operate the push-to-talk button and transmit a call in the following manner: "KOD-1590 calling KOD-1939." If the called station does not respond, try again later. To call a unit of the same station, make the call in the following manner: "KOD-1590 calling Unit One." Or, to make a call from a mobile unit to another mobile unit, both of which are covered by the same station license, use the following procedure: "KOD-1590 Unit One calling Unit Two."

What is the procedure for responding to a call?

Upon intercepting a call addressed to your station, transmit the following reply: "KOD-1939 back to KOD-1590, over." (Of course, use applicable call signs, not the examples given.) A mobile unit would respond to its base station in the following manner: "Unit One back to KOD-1590, over." A mobile unit responding to a call from another mobile unit, covered by the same station license, would respond in the following manner: "KOD-1590 Unit One back to Unit Two, over."

What is the procedure used for terminating an exchange of communications?

When completing an exchange of communications with another station, use the following procedure: "KOD-1590 off and clear with KOD-1939." To terminate a conversation with a unit of the same station, use the following procedure: "KOD-1590 off and clear with Unit One." To clear with another mobile unit when operating a mobile unit, both covered by the same station license, use the following procedure: "KOD-1590 Unit One off and clear with Unit Two."

What is meant by the *10-code*?

The 10-code was originated by the police departments using two-way radio. It is now used by many other users of two-way radio. Table 3-1 lists the basic 10-code used by CB operators. It is a convenient way to limit radio transmission time. For exam-

Table 3-1. The Basic 10-Code

Code	Meaning	Code	Meaning
10-1	Receiving poorly	**10-27**	Moving to channel ___
10-2	Signals good	**10-28**	Temporarily out of service
10-3	Stop transmitting. Channel in use by others	**10-29**	Back in service
10-4	OK—affirmative message received	**10-30**	Does not conform to rules and regulations
10-5	Relay message	**10-33**	Emergency traffic at this station
10-6	Operator busy, stand by	**10-34**	Trouble at this station
10-7	Out of service	**10-35**	Confidential information
10-8	In service—subject to call	**10-36**	Accident at ___
10-9	Repeat transmission, poor reception	**10-37**	Wrecker needed at ___
10-10	Transmission completed	**10-38**	Ambulance needed at ___
10-11	Speaking too rapidly	**10-39**	Convoy or escort
10-12	Officials or visitors present	**10-41**	Switch to channel ___
10-13	Advise road and weather conditions	**10-60**	What is your message number
10-14	Correct time	**10-63**	Net directed
10-15	Make a pick-up of ___ at ___	**10-64**	Net clear
10-16	Have picked up or have in possession	**10-65**	Net message assignment
10-17	Urgent (business)	**10-66**	Cancellation
10-18	Anything for us	**10-67**	Clear for net message
10-19	Nothing for you. Return to station	**10-68**	Repeat message
10-20	What is your location	**10-70**	Net message
10-21	Call me by land line	**10-71**	Proceed with transmission in sequence
10-22	Report in person to ___	**10-79**	Report progress of fire
10-23	Stand by	**10-84**	What is your telephone number
10-24	Finished with last assignment	**10-91**	Too weak, talk closer to mike
10-25	Do you have contact with ___	**10-92**	Too loud, farther from mike
10-26	Disregard last information	**10-93**	Frequency check
		10-94	Give a test with voice
		10-95	Give a test without voice
		10-99	Unable to receive you

ple, a station wanting to know the location of another station would ask: "What is your 10-20?" The other station would reply: "My 10-20 is 14th and Grand."

What are the APCO 10-signals?

They are the abbreviations used by most police departments. These are listed in Table 3-2. Cards designed to fit a car sun

Table 3-2. Associated Public Safety Communications Officers, Inc. Official Ten-Signal List

10-0	Caution	**10-34**	Riot
10-1	Unable copy___change location	**10-35**	Major crime alert
10-2	Signal good	**10-36**	Correct time
10-3	Stop transmitting	**10-37**	(Investigate) suspicious vehicle
10-4	Acknowledgment (OK)	**10-38**	Stopping suspicious vehicle
10-5	Relay	**10-39**	Urgent—use light, siren
10-6	Busy—unless urgent	**10-40**	Silent run—no light, siren
10-7	Out of service	**10-41**	Beginning tour of duty
10-8	In service	**10-42**	Ending tour of duty
10-9	Repeat	**10-43**	Information
10-10	Fight in progress	**10-44**	Permission to leave ______ for ____
10-11	Dog case	**10-45**	Animal carcass at ____
10-12	Stand by (stop)	**10-46**	Assist motorist
10-13	Weather—road report	**10-47**	Emergency road repair at ____
10-14	Prowler report	**10-48**	Traffic standard repair at __
10-15	Civil disturbance	**10-49**	Traffic light out at ____
10-16	Domestic problem	**10-50**	Accident (F, PI, PD)
10-17	Meet complainant	**10-51**	Wrecker needed
10-18	Quickly	**10-52**	Ambulance needed
10-19	Return to ____	**10-53**	Road blocked at ____
10-20	Location	**10-54**	Livestock on highway
10-21	Call ______ by telephone	**10-55**	Intoxicated driver
10-22	Disregard	**10-56**	Intoxicated pedestrian
10-23	Arrived at scene	**10-57**	Hit and run (F, PI, PD)
10-24	Assignment completed	**10-58**	Direct traffic
10-25	Report in person (meet)______	**10-59**	Convoy or escort
10-26	Detaining subject, expedite	**10-60**	Squad in vicinity
10-27	(Drivers) license information	**10-61**	Personnel in area
10-28	Vehicle registration information	**10-62**	Reply to message
10-29	Check for wanted	**10-63**	Prepare make written copy
10-30	Unnecessary use of radio	**10-64**	Message for local delivery
10-31	Crime in progress	**10-65**	Net message assignment
10-32	Man with gun	**10-66**	Message cancellation
10-33	EMERGENCY	**10-67**	Clear for net message

Table 3-2—Continued

10-68 Dispatch information	**10-84** If meeting _____ advise ETA
10-69 Message received	**10-85** Delayed due to _____
10-70 Fire alarm	**10-86** Officer/operator on duty
10-71 Advise nature of fire	**10-87** Pickup/distribute checks
10-72 Report progress on fire	**10-88** Present telephone # of _____
10-73 Smoke report	**10-89** Bomb threat
10-74 Negative	**10-90** Bank alarm at _____
10-75 In contact with _____	**10-91** Pick up prisoner/subject
10-76 En route _____	**10-92** Improperly parked vehicle
10-77 ETA(Estimated Time Arrival)	**10-93** Blockade
10-78 Need assistance	**10-94** Drag racing
10-79 Notify coroner	**10-95** Prisoner/subject in custody
10-80 Chase in progress	**10-96** Mental subject
10-81 Breatherlizer report	**10-97** Check (test) signal
10-82 Reserve lodging	**10-98** Prison/jail break
10-83 Work school xing at _____	**10-99** Wanted/stolen indicated

visor which list the APCO 10-signals on one side and other valuable data on the other side are available in sets of 30 cards for $2.00, from Associated Public Safety Officers, Inc., P.O. Box 669, New Smyrna Beach, Florida, 32069.

What are "Q" signals?

"Q" signals are used throughout the world by ships and other radio stations to exchange communications on a brief basis. Table 3-3 lists the most commonly used "Q" signals. Their use by CB stations is permitted by the FCC. For example, QSO means "communication with another station."

Table 3-3. Q Signals Adapted for Mobile Radiocommunications Applications

Signal	Question	Reply
QRA	What **station** are you?	I am **station** ______.
QRB	How **far** are you from me?	I am ______ miles away.
QRD	Where are you **headed** and from where?	I am bound for ______ from ______.
QRE	What is your estimated **time of arrival** at ______?	I expect to arrive in ______ at ______.
QRF	Are you **returning** to ______?	I am returning to ______.
QRG	What is my exact **frequency?**	Your frequency is ______.
QRK	How do you read my **signals?**	Your signals are ______. (1) Unreadable (2) Readable now and then (3) Readable with difficulty (4) Readable (5) Perfectly readable
QRL	Are you **busy?**	I am busy.
QRM	Are you experiencing **interference?**	I am experiencing interference.
QRN	Are you troubled by **static?**	I am troubled by static.
QRT	Shall I **stop** transmitting?	Stop transmitting.
QRU	Have you **anything** for me?	I have nothing for you.
QRV	Are you **ready?**	I am ready.
QSA	What is the strength of my **signals?**	Your signals are ______ (1) Scarcely perceptible (2) Weak (3) Fairly good (4) Good (5) Very good
QSB	Are my signals **fading?**	Your signals are fading.
QSL	Will you send me a **confirmation** of our communication?	I will confirm.
QSM	Shall I **repeat** the last message?	Repeat the last message.
QSO	Can you **communicate** with ______?	I can communicate with ______.
QTC	How many **messages** do you have for me?	I have ______ messages for you.
QTH	What is your **location?**	I am at ______.
QTN	At what **time** did you **depart** from ______?	I left ______ at ______.
QTO	Have you **left port** (dock)?	I have left port (dock).
QTP	Are you going to **enter port** (dock)?	I am going to enter port (dock).
QTR	What is the correct **time?**	The correct time is ______.

Table 3-3.—Continued

QTU	During what **hours** is your station open?	My station is open from ________ to ________.
QTV	Shall **I stand guard** for you on ________ MHz/kHz	Stand guard for me on ________ MHz/kHz.
QTX	Will you keep your station open for **further communication** with me for ________ hours?	I will keep my station open for further communication with you for ________ hours.
QUA	Do you have **news** of ________?	Here is news of ________.

Examples of use:
Question: QRF warehouse five?
Reply: QRF warehouse five.

Table 3-4. Phonetic Alphabet

Letter to be Identified	Identifying Word
A	Adam
B	Boy
C	Charles
D	David
E	Edward
F	Frank
G	George
H	Henry
I	Ida
J	John
K	King
L	Lincoln
M	Mary
N	Nora
O	Ocean
P	Paul
Q	Queen
R	Robert
S	Sam
T	Tom
U	Union
V	Victor
W	William
X	X-ray
Y	Young
Z	Zebra

When trying to identify a station by its call sign, it is often confusing to determine what letters and numbers are represented. Telephone operators, for example, try very hard to enunciate numbers and letters to make themselves understood. How is this done in radiocommunications?

Table 3-4 lists the phonetics used by most police departments for transmitting letters clearly. Hams often use phonetics of their own. For example, a ham station such as W6BCH would identify itself as "William Six Big California Ham." In CB operations, the FCC requires licensees to identify their stations very clearly. For example, KOD-1590 could not be "Cod fifteen ninety." It would have to be "Kay Oh Dee One Five Nine Oh."

What is meant by "24-hour" time and how is it used?

Airline pilots, the military, many police departments, and others refer to the hour on the 0-24 basis which starts at one minute after midnight as 0001 hours and ends at midnight as 2400 hours. To be more precise, on the air when saying 8:30 P.M., for example, that time can be transmitted as "twenty hundred thirty". Table 3-5 lists the 24-hour clock references to the 12-hour clock.

Are there any time limits on CB transmissions?

Exchanges of communication between stations (Class-D) covered by different station licenses are limited to five minutes. After conclusion of an interstation communication, no further

Table 3-5. 2400 Hour Time

2400	Midnight (twenty-four hundred)
0001	One minute after midnight (zero zero zero one)
0015	Quarter past midnight (zero zero one five)
0045	45 minutes past midnight (zero zero four five)
0100	One o'clock in the morning (zero one hundred)
0130	One thirty AM. (zero one three zero)
0200	2 AM (zero two hundred)
0300	3 AM
0400	4 AM
0500	5 AM
0600	6 AM
0700	7 AM
0800	8 AM
0900	9 AM
1000	10 AM (ten hundred)
1100	11 AM (eleven hundred)
1200	NOON
1201	One minute after noon (Twelve zero one)
1215	Quarter past noon (Twelve fifteen)
1300 (add 100 to 1200)	1 PM (Thirteen hundred)
1345 (add 045 to 1300)	1:45 PM (Thirteen forty-five)
1400 (add 200 to 1200)	2 PM
1500 (add 300 to 1200)	3 PM
1600 (add 400 to 1200)	4 PM
1700 (add 500 to 1200)	5 PM
1800 (add 600 to 1200)	6 PM
1900 (add 700 to 1200)	7 PM
2000 (add 800 to 1200)	8 PM (Twenty hundred)
2100 (add 900 to 1200)	9 PM (Twenty one hundred)
2200 (add 1000 to 1200)	10 PM
2300 (add 1100 to 1200)	11 PM

interstation communications are permitted until after a five-minute waiting period. However, communication between units of the same station are not subject to the above restrictions.

What are the limits for test transmissions?

Obviously, at the time of installation and when maintenance work is being performed, it is necessary to make test transmissions. No transmission should take place when the channel being used is not clear. Test transmissions must be limited to one minute in duration. It is unfair to make long test transmissions which can cause harmful interference to others who want to use CB for normal purposes.

Are CB radio stations allowed to relay messages?

No. CB radio stations are not supposed to accept messages for transmittal to third parties. However, there are exceptions. In cases of emergency, a CB radio operator may call the police or other public-safety organization to ask for assistance for another person.

May CBers report traffic information to broadcast stations for relay to the public?

Yes, when the CBer has entered into a written agreement with the broadcast station. This is a public service which should be encouraged.

Are CB stations permitted to communicate with amateur radio stations?

No, except in true emergencies, in which case a complete report must be filed with the FCC.

May a CB radio-station operator communicate with any other CB station over any distance?

No. Class-D stations may communicate only with other stations within 150 miles. Because of skip transmission, it is possible for CB stations to intercommunicate over distances of thousands of miles. But this is unlawful. The distance limits, however, do not apply to Class-A CB stations.

Does the 150-mile communication limit apply to the distance of a mobile unit from the address noted on the station license? In other words, does it mean that a mobile unit may not communicate with anyone when more than 150 miles from home base?

No. The 150-mile rule applies to the distance between two intercommunicating units. For example, when Mr. Smith has his mobile unit 500 miles from his registered station address, he may not lawfully communicate with his base station at that address, but he may communicate with other CB units within 150 miles.

What is meant by *talking skip?*

CBers who unlawfully communicate with other CBers more than 150 miles away are said to be "skip talkers".

How can reception of skip signals be prevented?

There is no known way to prevent reception of signals arriving at the antenna via skip propagation.

Is it lawful to repeat what has been overheard on a CB-station transmission?

No. The Communications Act of 1934, as amended, states that it is unlawful for anyone to reveal the contents of any transmission made by any class of radio station with the exception of Amateur and Broadcast. The law even prohibits revealing the existence of any such transmission.

When a CB operator says "break break," what does he mean?

He is trying to break into a conversation between other operators. This should not be done except in emergencies.

What does MAYDAY mean?

MAYDAY means the same as SOS. MAYDAY is used when employing radiotelephony, and SOS is used when employing radiotelegraphy. MAYDAY is a distress call which should never be used by anyone except when in dire emergency where the immediate safety of life and property are concerned. For example, MAYDAY should not be used to call for help as a result of an automobile accident or breakdown of a vehicle. It is better to use "10-33," which means "emergency traffic." When MAYDAY is heard by radio operators, it is customary for all other stations to stay off that channel and allow the emergency traffic to be handled in an orderly manner without interference. There should seldom if ever be a need for a CB station to use MAYDAY except perhaps on a boat that is in danger of sinking.

What is the CB emergency channel?

Channel 9 (27.065 MHz) has been officially designated by the FCC as an emergency channel in response to a petition filed by the late George Nims Raybin. The use of Channel 9 is restricted to communications of an emergency nature only. An example of the use of Channel 9 is shown in Fig. 3-6.

Since FCC rules restrict the use of Channel 9 to emergency traffic only, what operating procedures should be used by a motorist on a trip?

Set the transceiver to Channel 9 so that calls for help from other motorists can be intercepted. Then when a call is received, ac-

Fig. 3-6. The young lady can call for assistance on Channel 9.

knowledge the call on Channel 9 and ask the other station to switch over to another channel so that communications can be exchanged on the other channel. This will leave Channel 9 open for other emergency calls.

How can Channel 9 be monitored while listening on other channels?

A separate receiver tuned to Channel 9 can be used. Or a transceiver with built-in Channel 9 monitoring capability, such as the one shown in Fig. 3-7, can be used. Such a transceiver is always tuned to Channel 9, in addition to the channel selected with the channel selector. When a signal on Channel 9 is intercepted, a lamp glows, alerting the operator that a transmission on Channel 9 is taking place.

Courtesy Midland Communications Co.

Fig. 3-7. 23-Channel Class-D tranceiver with automatic Channel 9 monitoring feature.

On a boat equipped with only a CB radio, how can the aid of the Coast Guard be obtained?

Contact a CB station on shore and ask its operator to call the Coast Guard by land-line telephone. Be sure to give your location and information about the nature of your problem.

Why does the FCC prohibit CB operators from talking about the technical performance of their equipment on the air?

The purpose of the Citizens Radio Service is to provide the public with means for two-way radiocommunication for personal and business reasons. It is not intended that Citizens radio be used as a hobby in and of itself. Therefore, the FCC amended its rules to prohibit talking about the technical performance of the equipment, which has nothing to do with the actual benefits of using the equipment. It is unlawful to discuss the strength of the signal, the type of antenna used, etc., as hams do. Any person who wants to use radio as a hobby should become a ham.

Is it lawful for a person to send a QSL card to a CB station?

No, unless the CB station operator specifically states that he would like to have a card sent to him acknowledging receipt of his transmission or to confirm two-way radio contact. One way CB operators get around this rule is to have their names listed in a magazine in a column listing the names, addresses, and call signs of CB stations wishing to receive QSL cards. Fig. 3-8 shows a typical CB QSL card.

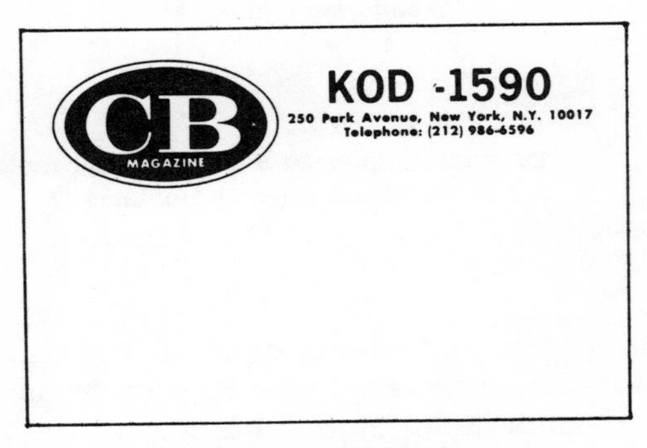

Fig. 3-8. Typical CB QSL card.

What should be done when some other CB operators are heard violating the rules and carrying on long, inane conversations?

If the identity of the rule violators is known, call them on the land-line telephone and suggest that they operate in a lawful manner. If their identities are not known, listen for their call signs and look up their names and addresses in a CB call book and then contact them to advise them to act like law-abiding human beings. The FCC does not encourage individuals to file complaints with them about the unlawful operating procedures used by other CB operators. The FCC would prefer to receive such complaints from an organized group such as a CB radio club.

What is a CB jamboree?

A CB jamboree is a social function where CBers gather for pleasure. It is strictly a voluntary activity that has no connection with the actual use of CB radio.

Are call books available which list the names and addresses of all CB licensees?

No, none at present. Lists of new CB licensees are available from CB Magazine, United Founders Tower, Oklahoma City, Oklahoma 73112.

Is it necessary to maintain a radio-station log?

No, the rules do not require CB radio stations to maintain a log. Nevertheless it is a good idea. Fig. 3-9 is an example of a log. Each transmission should be recorded. Whenever any repairs are made to the transceiver, this should also be noted in the log.

Date	Station Called	Called by Station	Location of Station	Time Started	Time Concluded	Subject Discussed	Channel Number

Fig. 3-9. Sample log-book page.

Is it lawful to monitor police radio transmissions?

Yes. Many CB club members and REACT groups monitor local police radio transmissions with a receiver such as the one shown in Fig. 3-10. Listening is not unlawful, but use of any information overheard is unlawful. Tow-truck operators, lawyers, etc., are known to have monitored transmission to learn of accidents, in order to be able to go to the scene and pick up some business. This is unlawful.

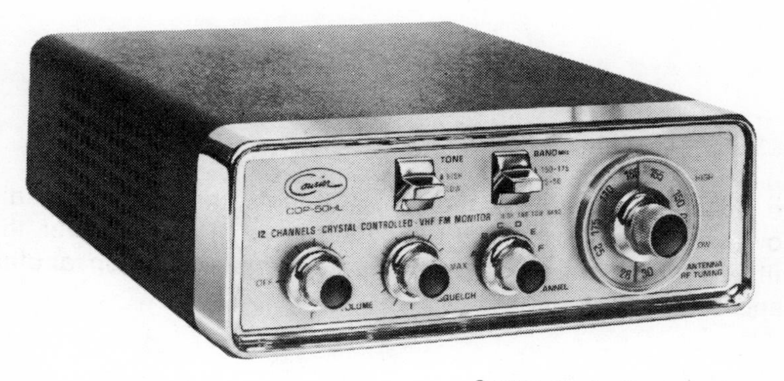

Courtesy Courier Communications

Fig. 3-10. Monitor receiver for the 25-50 MHz land mobile, 150-175 MHz land mobile, and marine bands.

Is it lawful to install a monitor receiver in a vehicle?

Yes, provided the monitor receiver cannot be tuned to police radio frequencies. In most states, such a receiver may be tuned to and used for listening to police radio transmissions without a special permit. In some states, including New Jersey and New York, a permit issued by the police is required.

What is a scanner receiver?

A scanner receiver is one that automatically tunes itself to preselected channels, one after the other. When it discovers a signal on a channel, it stops scanning and locks onto that signal so it can be heard. When the signal ceases, scanning resumes.

Can police calls be heard with a CB transceiver?

No, except those few police departments that use CB radio. However, a converter can be added to a CB transceiver that will enable reception of police radio transmission on another band.

How can a CBer assist law enforcement agencies?

Many CBers belong to clubs, search and rescue groups, and REACT teams which have made agreements with police authorities to assist the police in time of need. Many police departments welcome the assistance of responsible groups and individuals.

When a CB radio-equipped car is left at a garage or parking lot and an attendant uses the CB radio, who is responsible to the FCC in regard to the use of the equipment?

The licensee is responsible. While the use of the CB transceiver by an unauthorized person is unlawful, it is difficult to do anything about it. A degree of protection against such misuse of a CB transceiver can be provided by installing a key-lock switch in series with the lead that goes to the car battery. This can be readily installed by a radio service technician or auto electrician.

Is a U. S. citizen permitted to operate a CB transceiver in Canada?

Yes. It is necessary to obtain a Tourist Radio Permit from the Department of Transport in Canada. Application forms are available from the Telecommunications Branch, Department of Transport. Ottawa, Ontario, Canada.

Can a Canadian CBer operate a CB transceiver when in the United States?

Yes, provided he gets advance authority to do so from the FCC.

Can an American CBer use his CB transceiver in Mexico?

Yes, but only after receipt of authority from the government of Mexico. Information on how to obtain such authority can be had from the office of a Mexican consulate or from the Mexican Embassy in Washington, D.C.

If interference is received from diathermy and other machines operating within the Citizens band, what can be done about it?

Very little. The CB channels are available to Citizens-radio licensees with the understanding that there is no guarantee of immunity from interference caused by industrial, medical, and scientific apparatus permitted to operate within the Citizens band. However, if such devices do not meet FCC technical standards, the FCC can order that their use be ceased.

When transmitting with a CB transceiver causes interference to nearby television receivers, what should be done about it?

The trouble is often the fault of the television receiver. Many television receivers are designed to keep manufacturing costs low and therefore are more susceptible to interference. However, the owner of the television receiver will be annoyed when his reception is marred by a CB station. The interference can

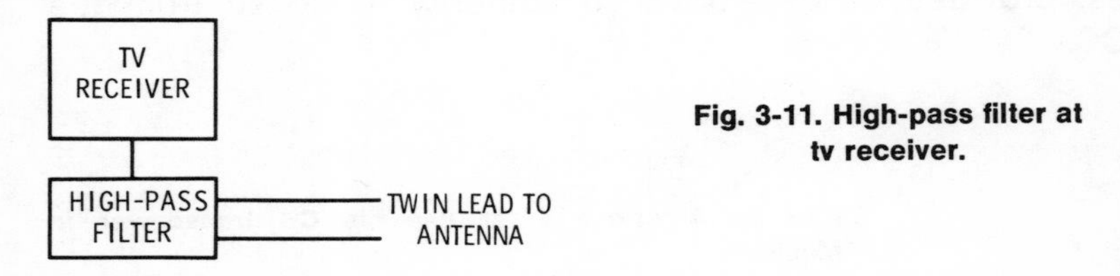

Fig. 3-11. High-pass filter at tv receiver.

usually be eliminated or at least minimized by installing a high-pass filter at the television-receiver antenna terminal, as shown in Fig. 3-11. If this does not stop the trouble, it may be necessary to install a low-pass filter between the CB transceiver and the coaxial cable leading to the antenna, as illustrated in Fig. 3-12. Both types of filters are available at radio parts distributors. An example of a low-pass filter is shown in Fig. 3-13.

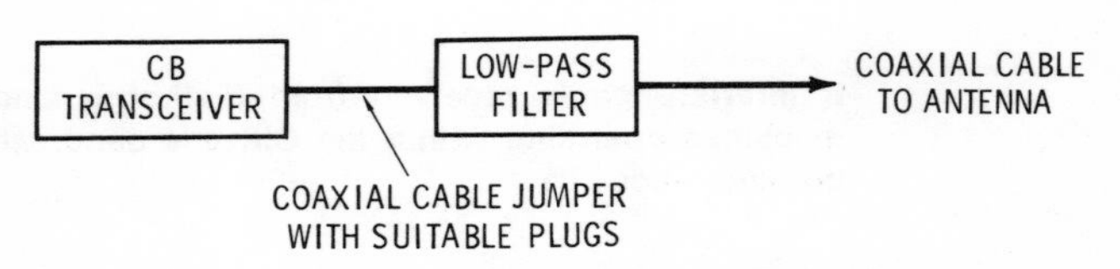

Fig. 3-12. Low-pass filter at CB transceiver.

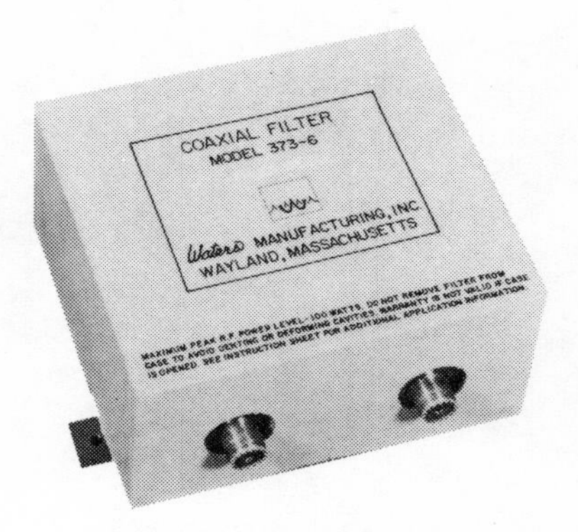

Courtesy Waters Manufacturing, Inc.

Fig. 3-13. Low-pass filter connected in series with the antenna transmission line. Used to attenuate harmonics that could cause TVI.

When a CB licensee receives a citation for off-frequency operation, what should be done?

Take the transceiver to a two-way radio repair shop and have the frequency measured and necessary corrections made. Get a written statement from the shop which makes it clear that the transceiver now meets FCC technical standards. Reply to the violation notice and include the statement from the repair shop.

4

CB Equipment

Where is CB radio equipment sold?

Nearly all of the nation's 2000 or so radio parts distributors sell CB equipment, as do thousands of CB dealers and numerous mail-order houses. Some stores handle several brands of equipment. Fig. 4-1 shows the CB display at the Paramus, New Jersey, branch of Lafayette Radio Electronics Corp.

What is the best way to select a CB transceiver for a particular application?

For general all-around use, a 23-channel transceiver (see Fig. 4-2) capable of operation from a 12-volt battery and 115 volts a-c, or from a 12-volt battery and through an adapter from 115 volts a-c, is recommended. However, lower-priced transceivers can be used for most applications. Not everyone needs access to all 23 channels. For example, a motorist who installs a CB transceiver in his car for the purpose of obtaining aid on the road, could get by with a single-channel transceiver. Nearly all

Courtesy Lafayette Radio Electronics Corp.

Fig. 4-1. Many electronics parts distributors handle several brands of CB equipment.

transceivers are capable of operation on five or more channels, but it is not necessary to equip the transceiver with crystals for all the channels.

Courtesy Pearce-Simpson, Inc.

Fig. 4-2. A 23-channel CB transceiver.

What criteria should be used when buying a CB transceiver?

An expert can read a spec sheet and make his own conclusions. The neophyte must depend upon the integrity of his dealer. The mere fact that the brand name is not well known is not an indication that the product is inferior. Some of the best transceivers are imports sold under many different labels.

What is meant by EIA standards?

Many equipment manufacturers build their CB equipment to conform with the technical standards of the Electronic Industries Association.

What is a solid-state transceiver?

One that employs no tubes. A solid-state transceiver uses transistors and other semiconductors in lieu of tubes.

Are tube-type CB transceivers obsolete?

No, they are not! Excellent tube-type transceivers are available. However, in mobile applications they draw more battery current than the solid-state types.

Does the number of transistors indicate the capability of a transceiver?

Not necessarily. For example, if a transceiver employs an IC (integrated circuit), the one IC might contain a half-dozen transistors which are not included in the transistor count. Also, it is possible to design a transceiver to use more transistors than actually required to obtain a specific sensitivity rating.

If 5 watts is the maximum legal input power of a Class-D transceiver, what is the output power?

The output power ranges from less than 3 watts to 4 watts, depending upon the efficiency of the transmitter. According to FCC rules, mean output power may not exceed 4 watts under any circumstances.

Can a separate transmitter and receiver be used in lieu of a transceiver?

Of course. The FCC only regulates the use of transmitters, whether independent or included in a transceiver, not the use of receivers. Separate CB transmitters and receivers are available. A crystal-controlled CB transmitter can be used in conjunction with a tunable short-wave receiver that tunes up to 30 MHz or higher.

Are CB equipment directories available?

Yes. Directories of CB equipment are available from Publishing Industries, Inc., United Founders Tower, Oklahoma City, Oklahoma 73112, and Davis Publications, 229 Park Ave. South, New York, New York 10003.

Can a CB transceiver be used on Business Radio Service channels?

Yes, if it has been type-accepted under Part 91, FCC Rules and Regulations. However, the transmitter must be licensed in the Business Radio Service. For use on CB channels, it must be licensed in the Citizens Radio Service. The same transmitter can be licensed in both radio services.

Can homemade equipment be used?

Yes, but it is seldom done. The transmitter would have to be considered by the FCC as acceptable for licensing. However, CB transceivers are available in kit form which can be assembled and wired at home. The transceiver shown in Fig. 4-3 is available in kit form.

Courtesy Allied Radio Shack

Fig. 4-3. Typical CB transceiver which is available in kit form.

What can be done with CB transceivers that are no longer to be used?

They can be donated to a school for use by students. Such donations can be made and the donor can claim a tax deduction for doing so. However, check this out with a tax accountant or the Internal Revenue Service for the latest rules.

What is the meaning of type acceptance?

When a transmitter has been type-accepted by the FCC, it means that the manufacturer has submitted documentary evidence to the FCC which certifies that the transmitter meets the technical standards established by the FCC. The FCC can withdraw type acceptance of any make or model of transmitter which it finds no longer meets applicable standards.

What is meant by receiver sensitivity?

It is the measure of a receiver's receptivity. For example, if the sensitivity of a receiver is 1 microvolt for 10 dB S/N, it means that a radio signal, only one-millionth of one volt in strength at its antenna terminal, will produce a sound in the loudspeaker which will contain ten times more signal power than noise power. (S/N means signal-to-noise ratio.) A receiver rated at 0.5 microvolt is even more sensitive; one rated at 2 microvolts is less sensitive.

What is meant by receiver selectivity?

Selectivity means the ability to reject radio signals on channels other than the channel to which the transceiver is set. When set to Channel 10, for example, and Channel 11 stations can be heard, this is called "bleed-over" by some CBers.

What is a selectivity filter?

It is a built-in device which improves the selectivity of a receiver. Technically, it is a bandpass filter and is usually of the mechanical, ceramic, crystal-lattice, or L-C type.

May CB stations use ssb?

Yes, Class-D CB stations may use either a-m (amplitude-modulation) or ssb (single-sideband) transmission. Ssb is far superior to a-m and provides greater range. A typical ssb CB transceiver is shown in Fig. 4-4.

Courtesy Midland International Corp.

Fig. 4-4. 23-channel combination am/ssb transceiver.

Is ssb better than a-m?

Yes. Using ssb, more effective use is made of the available power. When using a-m, only 25% of the power is utilized (the power in one of the sidebands). When using ssb, 100% of the power is in the transmitted sideband, since the carrier and the other sideband are suppressed in the transmitter.

Why does an ssb transceiver have a "clarifier" control?

In an ssb receiver, it is necessary to restore the carrier that was generated and suppressed in the distant transmitter. A beat frequency oscillator (BFO) in the receiver must be tuned carefully so that voices will be received clearly. This is done with the clarifier control.

Is it legal to use a Class-D ssb transceiver rated at 15 watts P.E.P.?

Yes. P.E.P. stands for "peak envelope power" which can exist without exceeding the 4-watt mean power output limit or the 5-watt average power input limit.

Is it possible to put the transceiver in the car trunk and operate it from the driver's position?

Yes. Some transceivers are designed for trunk installation and are provided with a remote-control head. Fig. 4-5 shows a telephone-type control unit.

Can a transceiver be installed temporarily in a car?

Yes. Many transceivers are designed so that they can be quickly attached and detached from a mounting bracket installed in a

Fig. 4-5. A telephone-type remote control unit for transceiver mounted in trunk.

car. Some transceivers, designed for operation only from a 12-volt battery, may be used at fixed locations by providing an a-c adapter. Some have a dual power supply which enables operation from either a 12-volt d-c or 115-volt a-c source. Fig. 4-6 shows a CB transceiver behind the seat of a sports car. It uses a plug-in, base-loaded antenna. The set can also be used at a fixed location with the plug-in antenna or an external antenna system.

Will a CB transceiver cause a vehicle battery to run down?

Any load on the battery will cause it to discharge, except when the engine is running and the battery is being recharged. How-

Photo by Jacques Saphier

Fig. 4-6. Universal-type transceiver shown mounted behind the seat of a sports car.

ever, when the engine is not running, the charge in the battery can be dissipated by leaving the CB transceiver turned on. A solid-state transceiver draws so little current that this is seldom a problem.

Can a desk-type microphone be used with a CB station?

Yes. Many makes and models of desk-stand microphones are available for use with CB transceivers. In some CB transceivers, the microphone cable runs directly into the transceiver, and its leads are soldered to the circuitry. Others are supplied with a microphone jack. This enables plugging in the microphone cord and using an alternate type of microphone. When selecting a microphone, it should have the same general characteristics as

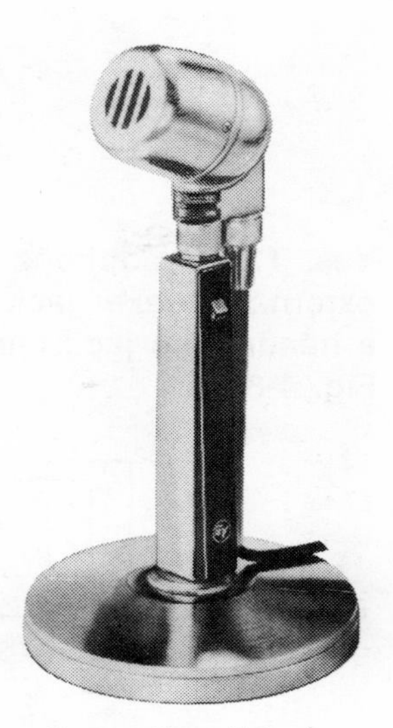

Fig. 4-7. Typical desk-stand microphone.

Courtesy Electro-Voice, Inc.

the one furnished with the transceiver. The CB radio dealer or radio parts distributor should be able to specify a microphone for use with your particular transceiver. A typical desk-stand microphone is shown in Fig. 4-7.

What is meant by PA capability?

Some transceivers have a PA (public address) switch and a terminal to which an external loudspeaker can be connected. When the switch is set to the PA position, the microphone, audio amplifier of the transceiver, and the external loudspeaker can be used as a public address system.

Can headphones be used with a CB transceiver?

Yes. The headphones (8-ohm type) can be plugged into the external speaker jack if one is provided. If one is not provided, a headphone jack can be added, and connected as shown in Fig. 4-8.

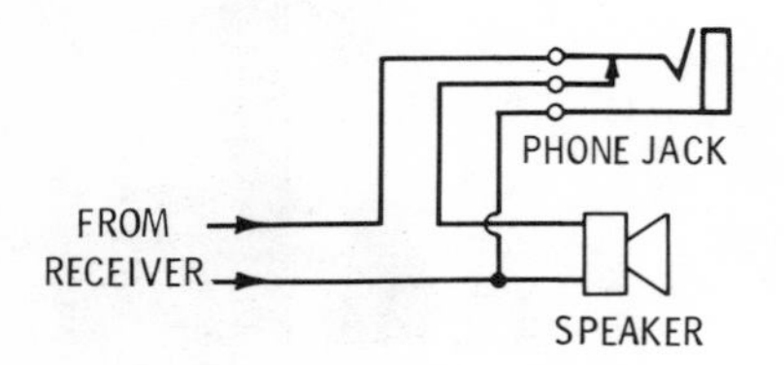

Fig. 4-8. Circuit for headphone jack. When the headphones are plugged in, the loudspeaker is silenced.

What is an "S" meter?

An "S" meter is a relative signal-strength meter wired into the receiver circuitry. It indicates the relative strength of the carrier of the radio signal being received. The meter (see Fig. 4-9) is usually calibrated in terms of "S" units from 1 to 9. A very strong signal would have a rating of S9 or higher. An "S" meter

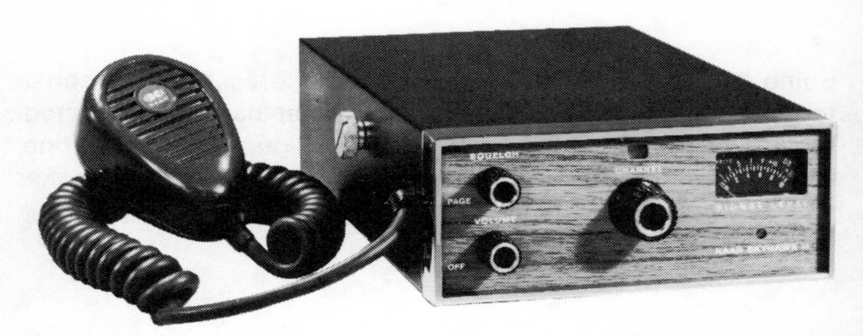

Courtesy Kaar Electronics Corp.

Fig. 4-9. Meter at upper right indicates relative strength of received signal.

has no practical value for communications purposes. It is a nice gadget to have, but does not contribute to the performance of the radio.

What is a noise limiter?

It is a built-in circuit in a receiver which reduces pulse-type noise such as ignition interference.

What is a noise blanker?

It is a circuit used in some receivers, in addition to a noise limiter, to further suppress pulse-type noise.

What is a noise silencer?

It can be a built-in noise blanker, or it can be an external device, such as the one made by Omega-T, which suppresses noise pulses before they reach the receiver.

What causes ignition interference?

An internal combustion engine generates radio-frequency energy in the form of putt-putt noise which can interfere with CB reception. This noise is caused by electrical discharges in the spark plugs and distributor and is radiated by the high voltage cables.

How can ignition noise be eliminated?

It might not be possible to eliminate ignition noise completely, but it can usually be reduced to an acceptable level by installing noise suppressors. In some stubborn cases it may be necessary to install an ignition shielding kit, such as the one shown in Fig. 4-10.

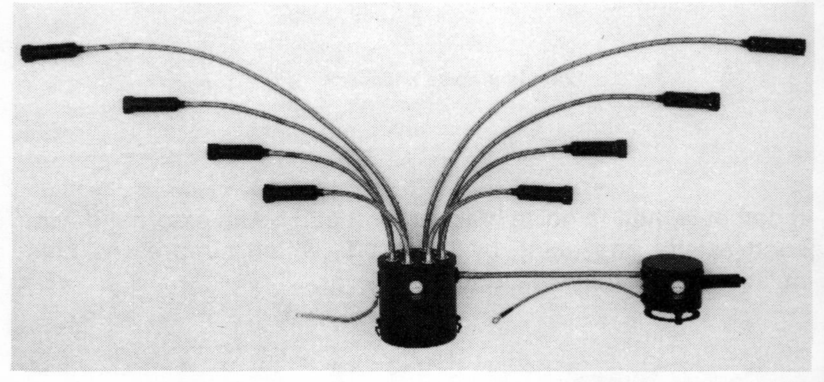

Courtesy Estes Engineering Co.

Fig. 4-10. Ignition shielding kit for suppressing radio interference.

What is an automatic transmitter identifier?

It is an encoder that automatically transmits a coded signal peculiar to that specific transceiver to enable the FCC to identify it with a decoder. Although not yet required, FCC officials have indicated that such devices might be required in the future.

Is it lawful to use a transmitter power booster?

No, if the power is increased above 4 watts mean power (output) in the case of a Class-D station, or above 44 watts in the case of a Class-A station. The use of the linear amplifier, shown in Fig. 4-11, by a Class-D station is unlawful. But a Class-A station may use the booster amplifier shown in Fig. 4-12.

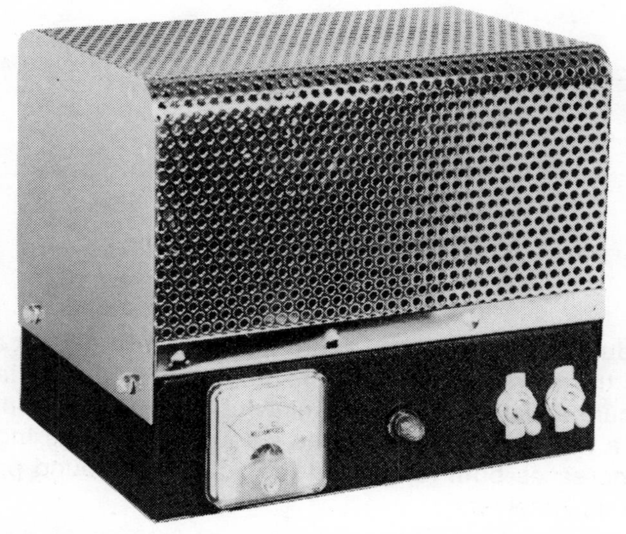

Fig. 4-11. A linear amplifier. Its use as a power booster by a Class-D station is unlawful.

Courtesy Kaar Electronics Corp.

Fig. 4-12. A booster amplifier which may be used with a Class-A station.

What is a modulation booster?

A modulation booster is an internal or external device which raises the output level of the microphone, but also limits the maximum level to prevent unlawful over-modulation. When using a modulation booster, the sideband power is increased. This increases both range and the volume of sound produced at distant receivers.

Is it permissible to use a phone patch?

Yes. The FCC does not prohibit the use of a phone patch. However, the local telephone company can make a monthly rental charge for an interface device. Some telephone companies will install one owned by the subscriber or will rent one to the subscriber. Fig. 4-13 shows how a telephone patch functions. When a telephone patch is used, it is possible for people at telephones at a distance to talk and listen over a CB station. There are limitations, however; the person authorized by the FCC to operate a CB station must physically control the transmitter. Also, it is a violation of FCC rules for a licensee to relay or transmit messages for anyone other than the licensee or his immediate family.

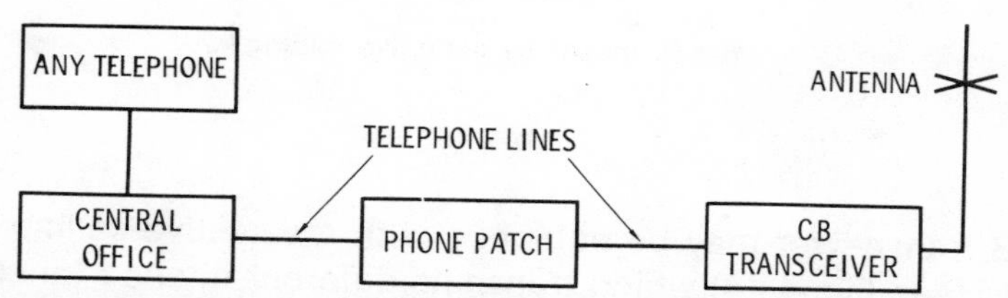

Fig. 4-13. Phone patch permits persons at a distance to communicate through a CB station.

What is a VOX?

VOX means "voice-operated switch" and enables voice-actuation of a transmitter, eliminating the need for a push button on the microphone.

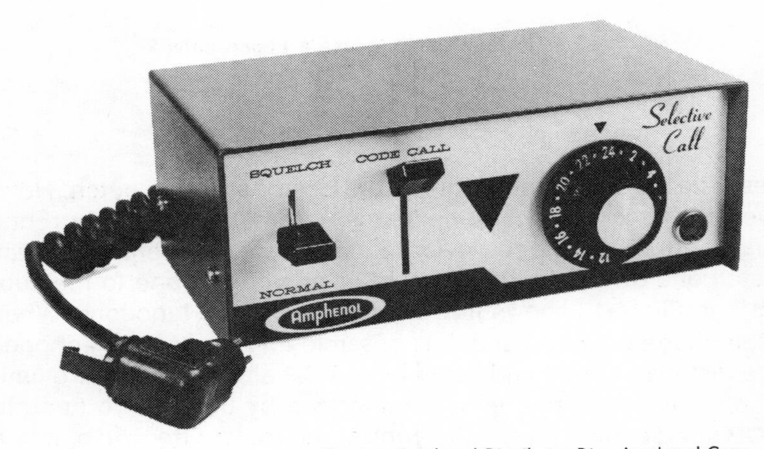

Courtesy Amphenol Distributor Div., Amphenol Corp.

Fig. 4-14. A selective-calling encoder.

What is meant by selective calling?

A CB transceiver may be equipped with a selective-calling encoder (see Fig. 4-14) which transmits different pitched tones or combinations of tones to signal other transceivers individually. These other transceivers are equipped with decoders which cause the speaker to be mute until a specific tone or combination of tones is intercepted. Thus, the users hear nothing until signaled.

What is meant by tone-coded squelch?

An ordinary CB transceiver will receive all transmissions on the channel to which it is set. One equipped with a tone-coded squelch decoder will respond only to transmissions preceded

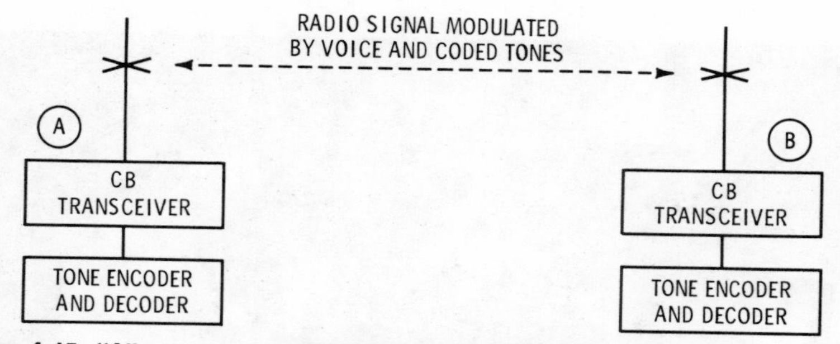

Fig. 4-15. "A" must transmit the correct coded-tone signal for reception by "B" and vice versa.

by, or accompanied by, a tone signal or combination of tones. Tone-coded squelch can be added to any CB transceiver when the user wants to limit reception to specific stations. Fig. 4-15 shows how tone-coded squelch functions.

Is it lawful to use a scrambler to make transmissions unintelligible?

Class-D stations may not use scramblers. Some other classes of stations may use them. (See the FCC rules.)

Can a licensee repair his own CB transceiver?

Yes, if he also possesses a First- or Second-Class Radiotelephone Operator License or does the work under the supervision of the holder of such a license. It is recommended that repairs be made only by a professional licensed technician who has the necessary test equipment. Fig. 4-16 shows a professional technician checking out a CB transceiver.

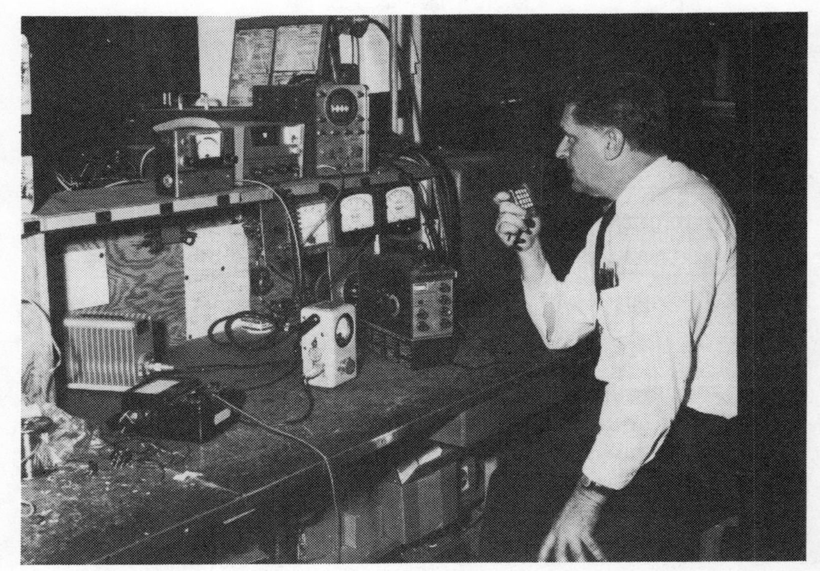

Courtesy The Hammarlund Mfg. Co.

Fig. 4-16. Only a licensed professional technician should service CB transceivers.

When a transceiver is purchased from a mail-order house and there is no local CB radio dealer, where can repair service be obtained when required?

The transceiver can, of course, be sent back to the company from which it was purchased. However, it will require considerable time for shipment and return. In most communities, there are one or more two-way radio repair shops which maintain commercial land mobile radio equipment and which are staffed by persons who can also service CB equipment. They are usually listed in the yellow pages of telephone directories under "Radiocommunications Equipment and Service." When one cannot be found in the telephone directory, contact the local police department. They undoubtedly know of local service shops. Also, licensed operators working at broadcasting stations are usually capable of servicing CB equipment.

How often should a CB transceiver be checked out by a licensed technician?

At least once a year to ensure that it still performs in accordance with FCC technical standards. Among other things, the transmit frequencies should be measured.

Where can CB transceiver servicing information be obtained?

Generally, a service manual is packed with a new transceiver. An additional copy can usually be obtained from the manufacturer. Service information on older models, including many that are no longer on the market, can be found in the CB Radio Series published by Howard W. Sams & Co., Inc. The various volumes of this series can be bought or ordered through any book dealer or Sams distributor.

If all of the channels for which a CB transmitter is equipped are excessively congested, what is the procedure for having crystals installed for other channels?

Take the transceiver to a two-way radio repair shop or the dealer from whom it was purchased and have a licensed technician install crystals for a new channel. At the same time, have the frequency checked to make sure that the new channels will be within the frequency tolerance established by the FCC. While many CB operators install crystals themselves, it is unlawful to do so.

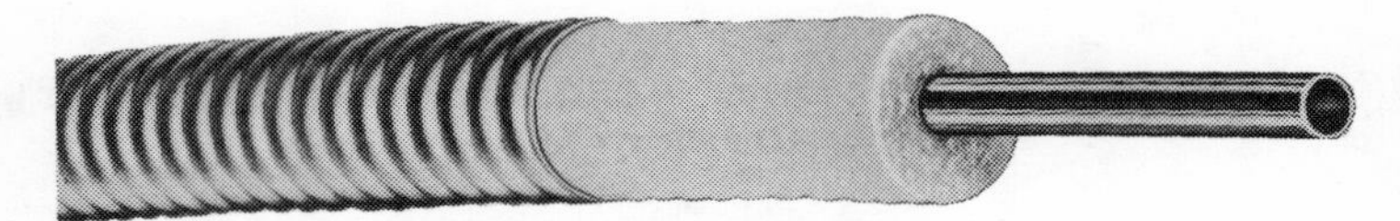

Fig. 4-17. Low-loss coaxial cable.

When the communicating range from a base station is inadequate, what can be done about it?

If the problem is caused by noise from passing motor vehicles, the antenna should be moved to a location farther away from the street. This will require a longer piece of coaxial cable for connection to the transceiver. But, by using low-loss coaxial cable such as shown in Fig. 4-17, the losses in the cable are minimized. FCC rules prohibit installation of the antenna so that its top protrudes more than 20 feet above a man-made structure, natural formation, or the ground. It is generally believed that maximum communicating range can be attained by installing the antenna as high as possible. This is true when the antenna is mounted on a fairly tall structure. When it is not possible to install the antenna on top of a tall structure, better results can sometimes be obtained by mounting the antenna right on the ground. This forms what is called a Marconi antenna, and experience has shown that it is quite effective, provided that the antenna is not surrounded by nearby metallic objects.

5

Mobile and Fixed Installations

Why should a solid-state CB transceiver be installed so it won't be in the direct hot air stream of a car heater?

Transistors are sensitive to heat. The hot air from the heater could affect transceiver performance and reduce transistor life.

What is meant by negative ground?

In a vehicle, either the negative or positive battery terminal is grounded to the frame. If a CB transceiver is designed for use only in a vehicle with negative ground, it cannot be used in a vehicle with positive ground. Some transceivers have a "floating" ground system and can be used in either type of vehicle.

How can a 12-volt transceiver be used in a vehicle with a 6-volt battery?

A 6-volt to 12-volt dc-to-dc power converter can be used, but they are expensive. Another way is to add a second 6-volt storage battery, a diode, and a DPDT knife switch, connected as shown in Fig. 5-1. When using the transceiver, the batteries are in series. At other times, the batteries are in parallel so both will be charged by the vehicle's generator. The diode prevents the auxiliary battery from discharging into the main battery.

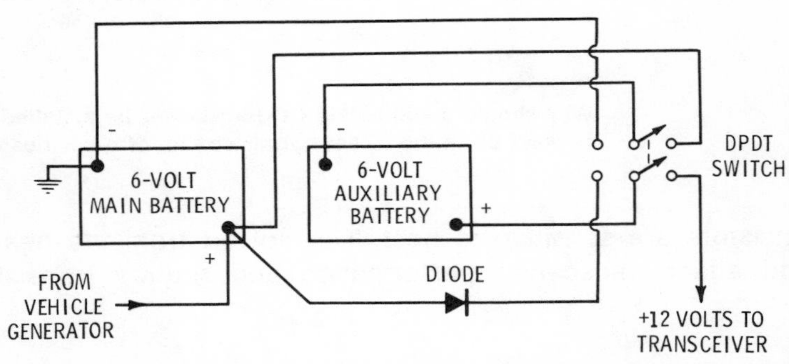

Fig. 5-1. Connections permitting 12-volt operation in a 6-volt vehicle.

How can a base station be connected so it will operate when there is a utility power failure?

If the transceiver is a 12-volt solid-state type, it can be operated from a 12-volt storage battery and a battery charger, connected as shown in Fig. 5-2. Normally, the charger furnishes power to the transceiver and keeps the battery charged. When utility power fails, the charge in the battery will keep the transceiver operational until the charge is exhausted or until utility power is restored.

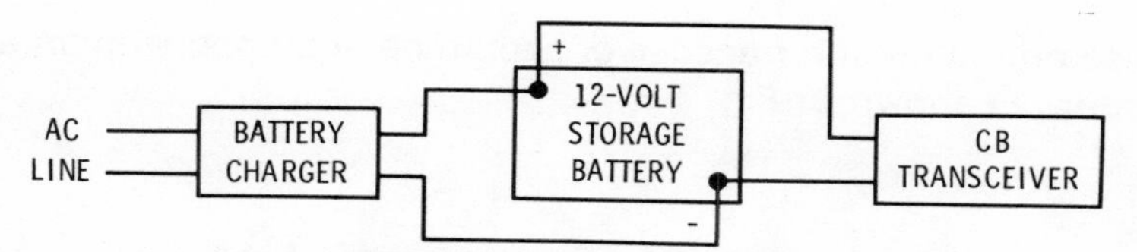

Fig. 5-2. Base stations operated from a battery and charger.

Is it lawful to operate a Class-D base station by remote control?

Yes, if the control point and the transceiver are on the same premises and they are interconnected by wires.

Can a Class-D CB transceiver be used for both voice communication and as a garage door opener?

Yes, if it is covered by both a Class-C and a Class-D station license. When used as a garage door opener, the transceiver is set to a Class-C channel (26.995, 27.095, 27.145, 27.195, or 27.255 MHz). The garage-door control receiver picks up this signal and actuates the garage-door control system. To prevent

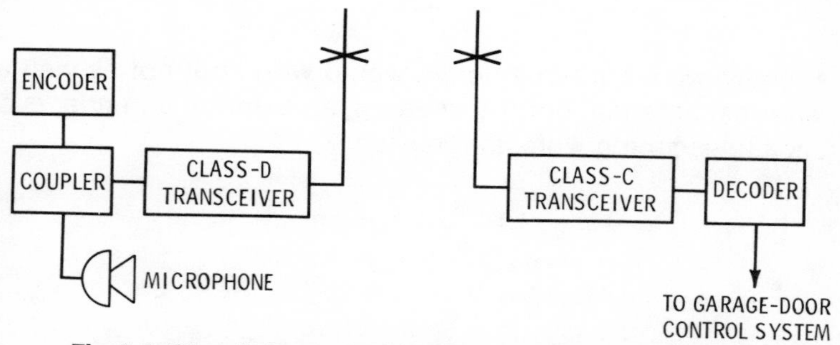

Fig. 5-3. Class-D transceiver utilized to control a garage door.

false operation, the transceiver should be equipped with an encoder, as shown in Fig. 5-3.

What is site noise?

It is the radio-frequency noise at a particular site. This noise can be generated by electrical devices or can exist because of natural phenomena.

Why is an antenna required?

The antenna radiates and intercepts the radio signals. To radiate effectively, the antenna must have an electrical length which is critical.

Why don't CB transceivers have built-in antennas such as are used in portable transistor radios?

For receiving, a built-in antenna would work, but not as well as an external antenna. For transmitting, an external antenna, even if it is a telescoping whip, is required.

How does antenna height affect range?

The higher the antenna is above the ground, the farther it is from the antenna to the horizon. Radio signals at frequencies above 25 MHz lose less of their energy when traveling over a line-of-sight path than when they are obstructed.

How is the antenna connected to the transceiver?

Except for hand-held walkie-talkies, coaxial cable is used.

Does it matter what kind of coaxial cable is used?

Yes indeed. The coaxial cable should have a characteristic impedance of 50 or 51 ohms. Type RG-58/U is satisfactory for runs up to 50 feet. For longer runs, type RG-8/U or other lower loss cable should be used.

What kinds of antennas are used at base stations?

Several types are available. The standard is the "ground plane" antenna which has a vertical radiator and three or four horizontal or drooping radials which form the ground plane. Less ob-

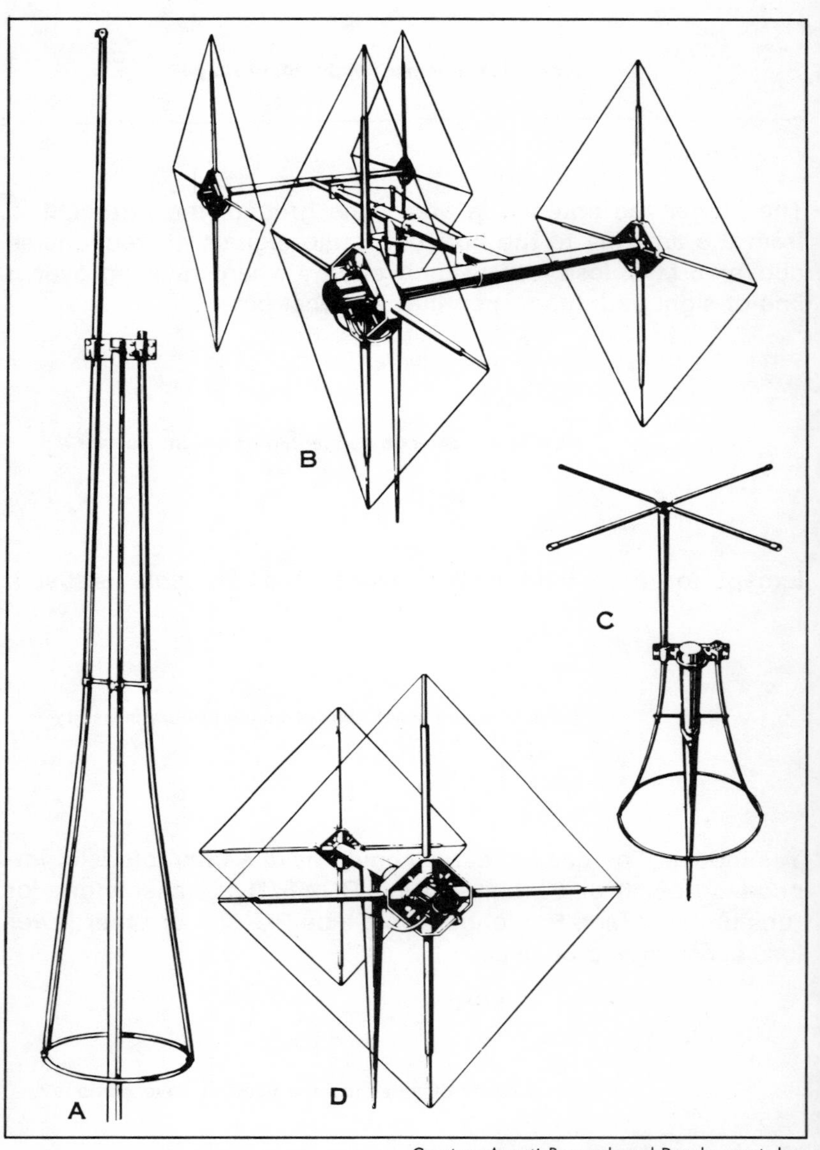

Courtesy Avanti Research and Development, Inc.

Fig. 5-4. Examples of base-station antennas. (A) Omnidirectional antenna. (B) Stacked PDL's switchable from horizontal to vertical mode. (C) Astrol Plane. Omnidirectional CB antenna. Out-performs all collinear and standard ground planes. (D) PDL. Switchable from horizontal to vertical operating mode.

trusive types do not have long radials or have none. Examples of base station antennas are shown in Fig. 5-4.

What kind of CB antenna can an apartment dweller use?

When the landlord will not permit installation of a CB antenna on the roof, a window-mount CB antenna can be used. Or a side-mount antenna can be used, as shown in Fig. 5-5.

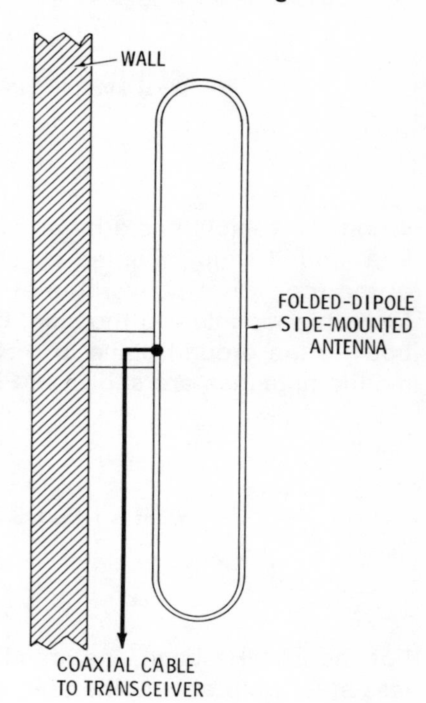

Fig. 5-5. Side-mount antenna for installation on the side of a building or tower.

Is it lawful to mount a CB antenna on the top of a tall television receiving antenna tower?

No, under existing Class-D rules, if the top of the CB antenna will be more than 20 feet above the ground. However, a CB antenna may be installed on the side of a transmitting antenna tower of any height provided the CB antenna does not extend above the tower, and provided the tower is used primarily by a station other than one licensed in the Citizens Radio Service.

What kind of antenna should be used on a vehicle?

When the vehicle has a metal roof, the most practical antenna is a loaded whip mounted as close as possible to the center of the roof. On a convertible or jeep, the antenna can be a 9-foot whip mounted to the side, bumper, or fender. The vehicle body is the ground plane of the antenna system. Examples of mobile antennas are shown in Fig. 5-6.

What is a loaded antenna?

For the 27-MHz band, the standard mobile antenna is a stainless-steel whip about nine feet long. The antenna can be made shorter physically while maintaining its electrical length by means of a loading coil at its base or in the middle. Fig. 5-7 shows an example of a loaded antenna.

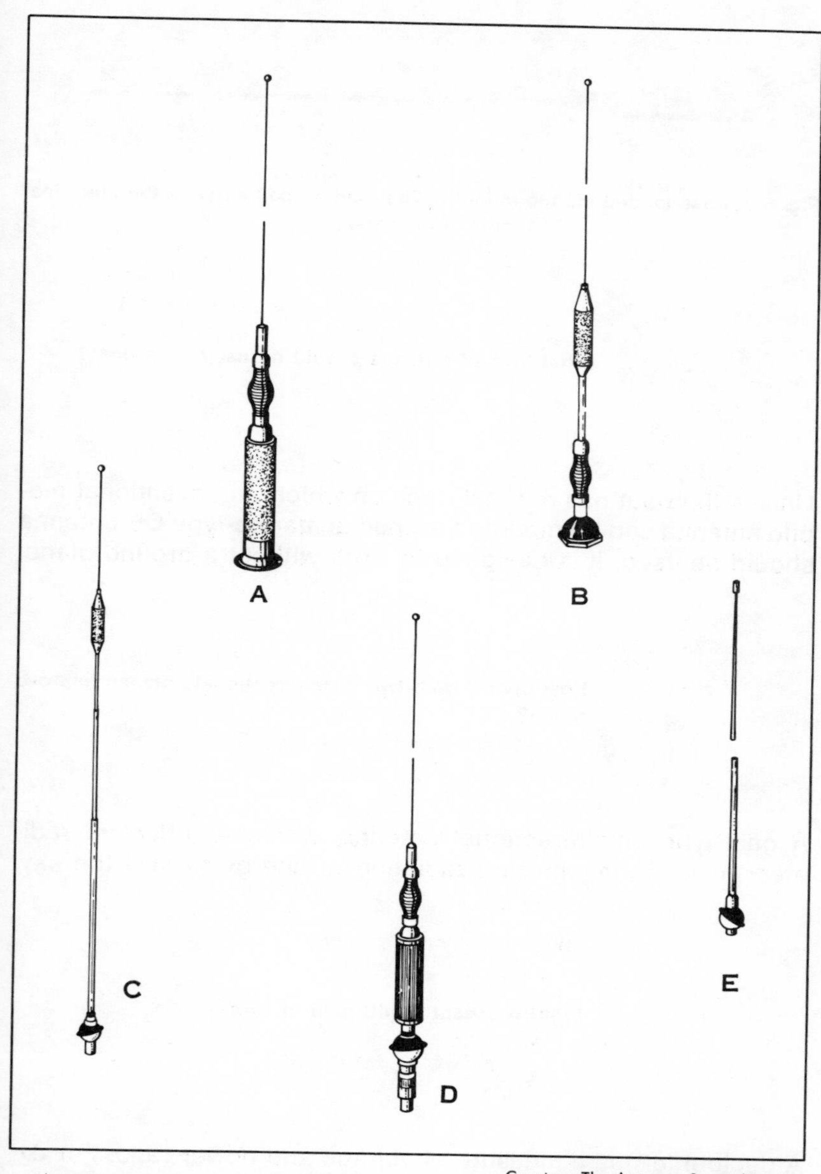

Courtesy The Antenna Specialists Co.

Fig. 5-6. Examples of mobile antennas. (A) Base-loaded antenna, only 46 inches long. (B) Roof-mounted antenna with waterproof snap-in mount and center-loading coil. (C) Center-loaded coil mount antenna. (D) Cowl, fender, or slanted-deck mount antenna. (E) Center-loaded fiberglass antenna.

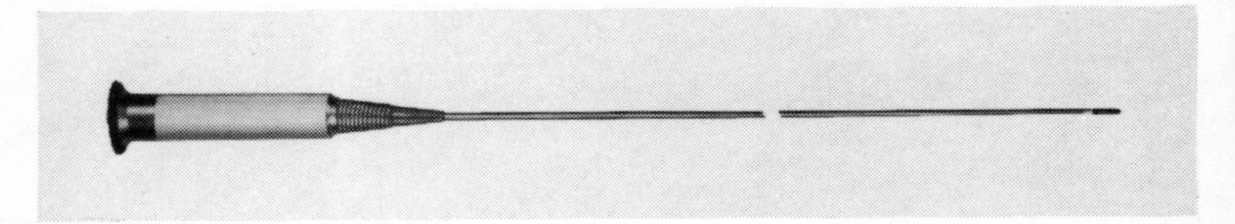

Fig. 5-7. Base-loaded mobile antenna. The loading coil extends the electrical length of the antenna.

What kind of antenna should be used on a boat?

Unless the boat has a metal deck on which a conventional mobile antenna can be mounted, a special marine-type CB antenna should be used. It is designed to work without a ground plane.

How does a gain-type omnidirectional antenna improve range?

A gain-type omnidirectional antenna increases effective radiated power by minimizing radiation of energy toward the sky.

What is meant by dB gain or loss?

A decibel (dB) is a measure of voltage and power ratio. For example, if an antenna provides 3 dB of gain, it has the same effect as doubling the power. If the antenna transmission line (coaxial cable) loss is 1 dB, the antenna system gain will be 2 dB (3 dB minus 1 dB). The total power gain will be 1.6 times.

What is a beam antenna?

A beam antenna radiates and receives signals better in one direction than in others. It also usually increases the effective radiated power (and effective sensitivity) in the favored direction.

When are horizontally polarized and vertically polarized antennas used?

Nearly all CB base stations and mobile units employ vertically polarized antennas. The electrostatic waves are vertical with respect to the earth. Horizontally polarized antennas are used primarily for point-to-point communication between fixed stations.

What is an antenna matcher?

It is an impedance matching device most often connected near the transceiver, as shown in Fig. 5-8 (A), to enable matching of the transceiver to the coaxial cable transmission line. However, an antenna matcher is more effective when connected at the antenna, as shown in (B), to enable matching the antenna to the transmission line. Since the antenna matcher would be exposed to the elements and hard to get at, it is seldom used where it would do the most good. Either way, an SWR meter should be used when adjusting the antenna matcher.

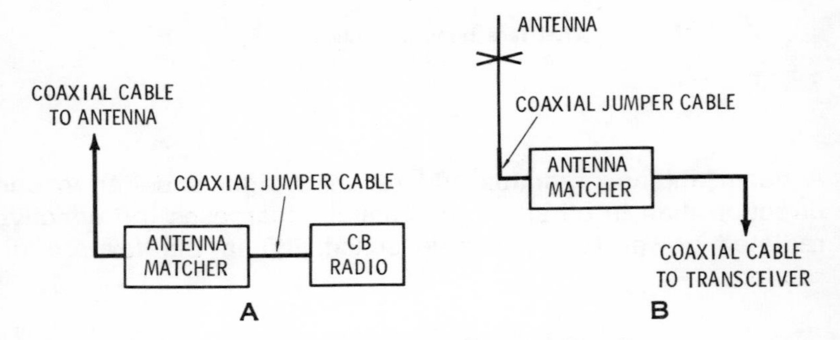

Fig. 5-8. Antenna matcher connections.

What is the difference between an antenna said to have a 1:1 SWR and one that has a rated SWR of 1.5:1?

If an antenna is claimed to have an SWR of 1:1, don't believe it, because it would be perfect and there is no such thing. One with an SWR rating of less than 2:1 is practical and quite efficient.

Is it necessary to tune the transceiver to the antenna system?

Yes indeed. The transceiver output trimmers should be adjusted so that the antenna system will absorb most of the energy and reflect as little energy as possible back to the transceiver. These adjustments should be made with an SWR (standing wave ratio) meter connected temporarily, as shown in Fig. 5-9, preferably by a licensed technician who knows what he is doing.

Fig. 5-9. Antenna-matching adjustments using SWR meter.

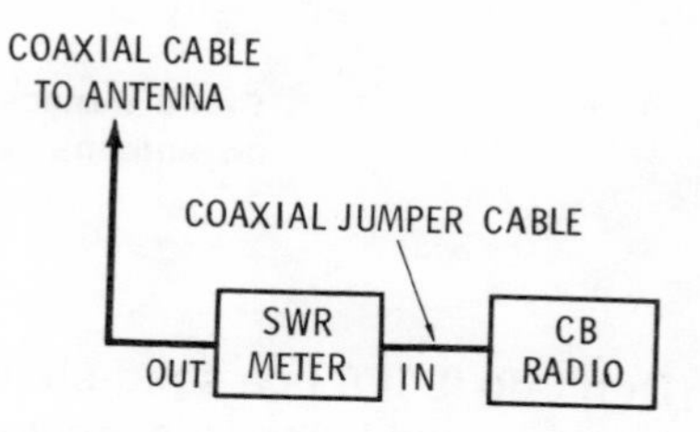

How can a base station be protected against lightning?

No lightning protection is foolproof. To discharge static accumulations around the antenna, the antenna supporting pipe should be grounded through a heavy copper wire, as shown in Fig. 5-10. Also, a coaxial-type lightning arrester should be connected in series with the antenna transmission line.

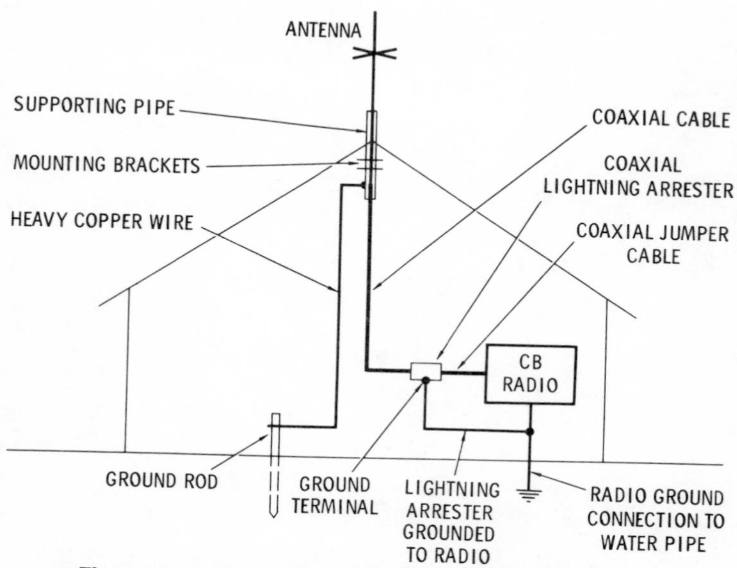

Fig. 5-10. Lightning protection system for a base station.

Can a transceiver be damaged by turning it on without an antenna connected?

Don't do it. If it is a solid-state type, one or more transistors can be damaged if the transceiver is set to transmit.

Index

D

E

F

G

H

I

L